Galton Francis

Natural Inheritance

Galton Francis

Natural Inheritance

ISBN/EAN: 9783337202248

Printed in Europe, USA, Canada, Australia, Japan

Cover: Foto ©berggeist007 / pixelio.de

More available books at **www.hansebooks.com**

NATURAL INHERITANCE

BY

FRANCIS GALTON, F.R.S.

AUTHOR OF

"HEREDITARY GENIUS," "INQUIRIES INTO HUMAN FACULTY," ETC.

CONTENTS.

CHAPTER V.

CHAPTER VI.

CHAPTER VII.

CHAPTER VIII.

CONTENTS.

CHAPTER IX.

CHAPTER X.

CHAPTER XI.

CHAPTER XII.

TABLES.

The words by which the various Tables are here described, have been chosen for the sake of quick reference ; they are often not identical with those used in their actual headings.

CONTENTS.

APPENDICES.

NATURAL INHERITANCE

NATURAL INHERITANCE.

CHAPTER I.

INTRODUCTORY.

I HAVE long been engaged upon certain problems that lie at the base of the science of heredity, and during several years have published technical memoirs concerning them, a list of which is given in Appendix A. This volume contains the more important of the results, set forth in an orderly way, with more completeness than has hitherto been possible, together with a large amount of new matter.

The inquiry relates to the inheritance of moderately exceptional qualities by brotherhoods and multitudes rather than by individuals, and it is carried on by more refined and searching methods than those usually employed in hereditary inquiries.

One of the problems to be dealt with refers to the curious regularity commonly observed in the statistical peculiarities of great populations during a long series of

B

generations. The large do not always beget the large,
nor the small the small, and yet the observed proportions between the large and the small in each degree of
size and in every quality, hardly varies from one generation to another.

A second problem regards the average share contributed to the personal features of the offspring by each
ancestor severally. Though one half of every child
may be said to be derived from either parent, yet he
may receive a heritage from a distant progenitor that
neither of his parents possessed as *personal* characteristics. Therefore the child does not on the average
receive so much as one half of his *personal* qualities
from each parent, but something less than a half. The
question I have to solve, in a reasonable and not merely
in a statistical way, is, how much less ?

The last of the problems that I need mention now,
concerns the nearness of kinship in different degrees.
We are all agreed that a brother is nearer akin than a
nephew, and a nephew than a cousin, and so on, but
how much nearer are they in the precise language of
numerical statement ?

These and many other problems are all fundamentally
connected, and I have worked them out to a first degree
of approximation, with some completeness. The conclusions cannot however be intelligibly presented in
an introductory chapter. They depend on ideas that
must first be well comprehended, and which are now
novel to the large majority of readers and unfamiliar
to all. But those who care to brace themselves to a

sustained effort, need not feel much regret that the
road to be travelled over is indirect, and does not
admit of being mapped beforehand in a way they can
clearly understand. It is full of interest of its own.
It familiarizes us with the measurement of variability,
and with curious laws of chance that apply to a vast
diversity of social subjects. This part of the inquiry
may be said to run along a road on a high level,
that affords wide views in unexpected directions, and
from which easy descents may be made to totally
different goals to those we have now to reach. I have
a 'great subject to write upon, but feel keenly my
literary incapacity to make it easily intelligible without
sacrificing accuracy and thoroughness.

CHAPTER II.

Natural and Acquired Peculiarities.—Transmutation of Female into Male Measures.—Particulate Inheritance.—Family Likeness and Individual Variation.—Latent Characteristics.—Heritages that Blend and those that are Mutually Exclusive.—Inheritance of Acquired Faculties.—Variety of Petty Influences.

A CONCISE account of the chief processes in heredity will be given in this chapter, partly to serve as a reminder to those to whom the works of Darwin especially, and of other writers on the subject, are not familiar, but principally for the sake of presenting them under an aspect that best justifies the methods of investigation about to be employed.

Natural and Acquired Peculiarities.—The peculiarities of men may be roughly sorted into those that are natural and those that are acquired. It is of the former that I am about to speak in this book. They are noticeable in every direction, but are nowhere so remarkable as in those twins[1] who have been dissimilar

[1] See *Human Faculty*, 237.

in features and disposition from their earliest years, though brought into the world under the same conditions and subsequently nurtured in an almost identical manner. It may be that some natural peculiarity does not appear till late in life, and yet may justly deserve to be considered natúral, for if it is decidedly exceptional in its character its origin could hardly be ascribed to the effects of nurture. If it was also possessed by some ancestor, it must be considered to be hereditary as well. But "Natural" is an unfortunate word for our purpose; it implies that the moment of birth is the earliest date from which the effects of surrounding conditions are to be reckoned, although nurture begins much earlier than that. I therefore must ask that the word "Natural" should not be construed too literally, any more than the analogous phrases of inborn, congenital, and innate. This convenient laxity of expression for the sake of avoiding a pedantic periphrase need not be accompanied by any laxity of idea.

Transmutation of Female into Male Measures.—We shall have to deal with the hereditary influence of parents over their offspring, although the characteristics of the two sexes are so different that it may seem impossible to speak of both in the same terms. The phrase of " Average Stature " may be applied to two men without fear of mistake in its interpretation; neither can there be any mistake when it is applied to two women, but what meaning can we attach to the word " Average " when it is applied to the stature of two such different

beings as the Father and the Mother? How can we
appraise the hereditary contributions of different ancestors whether in this or in any other quality, unless
we take into account the sex of each ancestor, in addition to his or her characteristics? Again, the same
group of progenitors transmits qualities in different
measure to the sons and to the daughters; the sons
being on the whole, by virtue of their sex, stronger,
taller, hardier, less emotional, and so forth, than the
daughters. A serious complexity due to sexual differences seems to await us at every step when investigating
the problems of heredity. Fortunately we are able to
evade it altogether by using an artifice at the outset, else,
looking back as I now can, from the stage which the
reader will reach when he finishes this book, I hardly
know how we should have succeeded in making a
fair start. The artifice is never to deal with female
measures as they are observed, but always to employ
their male equivalents in the place of them. I transmute all the observations of females before taking
them in hand, and thenceforward am able to deal
with them on equal terms with the observed male
values. For example: the statures of women bear to
those of men the proportion of about twelve to thir-
teen. Consequently by adding to each observed female
stature at the rate of one inch for every foot, we are
enabled to compare their statures so increased and transmuted, with the observed statures of males, on equal
terms. If the observed stature of a woman is 5 feet,
it will count by this rule as 5 feet + 5 inches; if it be

6 feet, as 6 feet + 6 inches; if $5\frac{1}{2}$ feet, as $5\frac{1}{2}$ feet + $5\frac{1}{2}$ inches; that is to say, as 5 feet + $11\frac{1}{2}$ inches.[1]

Similarly as regards sons and daughters; whatever may be observed or concluded concerning daughters will, if transmuted, be held true as regarding sons, and whatever is said concerning sons, will if re-transmuted, be held true for daughters. We shall see further on that it is easy to apply this principle to all measurable qualities.

Particulate Inheritance.—All living beings are individuals in one aspect and composite in another. They are stable fabrics of an inconceivably large number of cells, each of which has in some sense a separate life of its own, and which have been combined under influences that are the subjects of much speculation, but are as yet little understood. We seem to inherit bit by bit, this element from one progenitor that from another, under conditions that will be more clearly expressed as we proceed, while the several bits are themselves liable to some small change during the process of transmission. Inheritance may therefore be described as largely if not wholly "particulate," and as such it will be treated in these pages. Though this word is good English and accurately expresses its own meaning, the application

---

[1] The proportion I use is as 100 to 108; that is, I multiply every female measure by 108, which is a very easy operation to those who possess that most useful book to statisticians, *Crelle's Tables* (G. Reimer, Berlin, 1875). It gives the products of all numbers under 1000, each into each; so by referring to the column headed 108, the transmuted values of the female statures can be read off at once.

now made of it will be better understood through an illustration. Thus, many of the modern buildings in Italy are historically known to have been built out of the pillaged structures of older days. Here we may observe a column or a lintel serving the same purpose for a second time, and perhaps bearing an inscription that testifies to its origin, while as to the other stones, though the mason may have chipped them here and there, and altered their shapes a little, few, if any, came direct from the quarry. This simile gives a rude though true idea of the exact meaning of Particulate Inheritance, namely, that each piece of the new structure is derived from a corresponding piece of some older one, as a lintel was derived from a lintel, a column from a column, a piece of wall from a piece of wall.

I will pursue this rough simile just one step further, which is as much as it will bear. Suppose we were building a house with second-hand materials carted from a dealer's yard, we should often find considerable portions of the same old houses to be still grouped together. Materials derived from various structures might have been moved and much shuffled together in the yard, yet pieces from the same source would frequently remain in juxtaposition and it may be entangled. They would lie side by side ready to be carted away at the same time and to be re-erected together anew. So in the process of transmission by inheritance, elements derived from the same ancestor are apt to appear in large groups, just as if they had clung together in the pre-embryonic stage, as perhaps they did. They form what is well expressed by the word " traits," traits of feature and character—that is to say, continuous features and not isolated points.

We appear, then, to be severally built up out of a host of minute particles of whose nature we know nothing, any one of which may be derived from any one progenitor, but which are usually transmitted in aggregates, considerable groups being derived from the same progenitor. It would seem that while the embryo is developing itself, the particles more or less qualified for each new post wait as it were in competition, to obtain it. Also that the particle that succeeds, must owe its success partly to accident of position and partly to being better qualified than any equally well placed competitor to gain a lodgment. Thus the step by step development of the embryo cannot fail to be influenced by an incalculable number of small and mostly unknown circumstances.

Family Likeness and Individual Variation.—Natural peculiarities are apparently due to two broadly different causes, the one is Family Likeness and the other is Individual Variation. They seem to be fundamentally opposed, and to require independent discussion, but this is not the case altogether, nor indeed in the greater part. It will soon be understood how the conditions that produce a general resemblance between the offspring and their parents, must at the same time give rise to a considerable amount of individual differences. Therefore I need not discuss Family Likeness and Individual Variation under separate heads, but as different effects of the same underlying causes.

The origin of these and other prominent processes in heredity is best explained by illustrations. That which will be used was suggested by those miniature gardens, self-made and self-sown, that may be seen in crevices or other receptacles for drifted earth, on the otherwise bare faces of quarries and cliffs. I have frequently studied them through an opera glass, and have occasionally clambered up to compare more closely their respective vegetations. Let us then suppose the aspect of the vegetation, not of one of these detached little gardens, but of a particular island of substantial size, to represent the features, bodily and mental, of some particular parent. Imagine two such islands floated far away to a desolate sea, and anchored near together, to represent the two parents. Next imagine a number of islets, each constructed of earth that was wholly destitute of seeds, to be reared near to them. Seeds from both of the islands will gradually make their way to the islets through the agency of winds, currents, and birds. Vegetation will spring up, and when the islets are covered with it, their several aspects will represent the features of the several children. It is almost impossible that the seeds could ever be distributed equally among the islets, and there must be slight differences between them in exposure and other conditions, corresponding to differences in pre-natal circumstances. All of these would have some influence upon the vegetation; hence there would be a corresponding variety in the results.   In some islets one plant would prevail, in others another; nevertheless there would be many traits of family likeness in the vegetation of all of them, and no plant would be found that had not existed in one or other of the islands.

Though family likeness and individual variations arc largely due to a common cause, some variations are so large and otherwise remarkable, that they seem to belong to a different class.   They are known among breeders as "sports"; I will speak of these later on.

Latent Characteristics.—Another fact in heredity may also be illustrated by the islands and islets; namely, that the child often resembles an ancestor in some feature or character that neither of his parents personally possessed.   We are told that buried seeds may lie dormant for many years, so that when a plot of ground that was formerly cultivated is again deeply dug into and upturned, plants that had not been known to grow on the spot within the memory of man, will frequently make their appearance.   It is easy to imagine that some of these dormant seeds should find their way to an islet, through currents that undermined the island cliffs and drifted away their *débris*, after the cliffs had tumbled into the sea.   Again, many plants on the islands may maintain an obscure existence, being hidden and half smothered by successful rivals; but whenever their seeds happened to find their way to any one of the islets, while those of their rivals did not, they would sprout freely and assert themselves.   This illustration partly covers the analogous fact of diseases and other inheritances skipping a generation, which by the way I find to be by no means so usual an occurrence as seems popularly to be imagined.

Heritages that Blend and those that are Mutually Exclusive.—As regards heritages that blend in the offspring, let us take the case of human skin colour. The children of the white and the negro are of a blended tint ; they are neither wholly white nor wholly black, neither are they piebald, but of a fairly uniform mulatto brown. The quadroon child of the mulatto and the white has a quarter tint; some of the children may be altogether darker or lighter than the rest, but they are not piebald. Skin-colour is therefore a good example of what I call blended inheritance. It need be none the less " particulate " in its origin, but the result may be regarded as a fine mosaic too minute for its elements to be distinguished in a general view.

Next as regards heritages that come altogether from one progenitor to the exclusion of the rest. Eye-colour is a fairly good illustration of this, the children of a light-eyed and of a dark-eyed parent being much more apt to take their eye-colours after the one or the other than to have intermediate and blended tints.

There are probably no heritages that perfectly blend or that absolutely exclude one another, but all heritages have a tendency in one or the other direction, and the tendency is often a very strong one. This is paralleled

by what we may see in plots of wild vegetation, where
two varieties of a plant mix freely, and the general
aspect of the vegetation becomes a blend of the two,
or where individuals of one variety congregate and take
exclusive possession of one place, and those of another
variety congregate in another.

A peculiar interest attaches itself to mutually exclusive heritages, owing to the aid they must afford to the
establishment of incipient races.   A solitary peculiarity
that blended freely with the characteristics of the parent
stock, would disappear in hereditary transmission, as
quickly as the white tint imported by a solitary European would disappear in a black population. If the
European mated at all, his spouse must be black, and
therefore in the very first generation the offspring
would be mulattoes, and half of his whiteness would
be lost to them.   If these mulattoes did not inter-
breed, the whiteness would be reduced in the second
generation to one quarter ; in a very few more generations all recognizable trace of it would have gone.
But if the whiteness refused to blend with the black-
ness, some of the offspring of the white man would be
wholly white and the rest wholly black.   The same
event would occur in the grandchildren, mostly but
not exclusively in the children of the white offspring,
and so on in subsequent generations.   Therefore,
unless the white stock became wholly extinct, some
undiluted specimens of it would make their appearance during an indefinite time, giving it repeated

chances of holding its own in the struggle for existence,
and of establishing itself if its qualities were superior
to those of the black stock under any one of many
different conditions.

Inheritance of Acquired Faculties.—I am unpre-
pared to say more than a few words on the obscure,
unsettled, and much discussed subject of the possibility
of transmitting acquired faculties. The main evidence
in its favour is the gradual change of the instincts of
races at large, in conformity with changed habits, and
through their increased adaptation to their surroundings,
otherwise apparently than through the influence of
Natural Selection. There is very little direct evidence
of its influence in the course of a single generation, if
the phrase of Acquired Faculties is used in perfect
strictness and all inheritance is excluded that could be
referred to some form of Natural Selection, or of
Infection before birth, or of peculiarities of Nurture
and Rearing. Moreover, a large deduction from the
collection of rare cases must be made on the ground
of their being accidental coincidences. When this
is done, the remaining instances of acquired disease
or faculty, or of any mutilation being transmitted from
parent to child, are very few. Some apparent evidence
of a positive kind, that was formerly relied upon, has
been since found capable of being interpreted in another
way, and is no longer adduced. On the other hand there
exists such a vast mass of distinctly negative evidence,
that every instance offered to prove the transmission

of acquired faculties requires to be closely criticized.
For example, a woman who was sober becomes a
drunkard. Her children born during the period of her
sobriety are said to be quite healthy ; her subsequent children are said to be neurotic. The objections to accepting
this as a valid instance in point are many.   The woman's
tissues must have been drenched with alcohol, and the
unborn infant alcoholised during all its existence in that
state. The quality of the mother's milk would be bad.
The surroundings of a home under the charge of a
drunken woman would be prejudicial to the health of
a growing child. No wonder that it became neurotic.
Again, a large number of diseases are conveyed by
germs capable of passing from the tissues of the
mother into those of the unborn child otherwise than
through the blood.   Moreover it must be recollected
that the connection between the unborn child and the
mother is hardly more intimate than that between some
parasites and the animals on which they live.   Not
a single nerve has been traced between them, not a
drop of blood [1] has been found to pass from the mother
to the child.   The unborn child together with the
growth to which it is attached, and which is afterwards
thrown off, have their own vascular system to themselves, entirely independent of that of the mother.
If in an anatomical preparation the veins of the mother
are injected with a coloured fluid, none of it enters the
veins of the child ; conversely, if the veins of the child

[1] See *Lectures* by William O. Priestley, M.D. (Churchill, London, 1860),
pp. 50, 52, 55, 59, and 64.

are injected, none of the fluid enters those of the
mother. Again, not only is the unborn child a separate animal from its mother, that obtains its air and
nourishment from her purely through soakage, but its
constituent elements are of very much less recent
growth than is popularly supposed. The ovary of
the mother is as old as the mother herself; it was well
developed in her own embryonic state. The ova it contains in her adult life were actually or potentially present
before she was born, and they grew as she grew. There
is more reason to look on them as collateral with the
mother, than as parts of the mother. The same may
be said with little reservation concerning the male
elements. It is therefore extremely difficult to see
how acquired faculties can be inherited by the children.
It would be less difficult to conceive of their inheritance
by the grandchildren. Well devised experiment into
the limits of the power of inheriting acquired faculties
and mutilations, whether in plants or animals, is one of
the present desiderata in hereditary science. Fortunately
for us, our ignorance of the subject will not introduce
any special difficulty in the inquiry on which we are
now engaged.

Variety of Petty Influences.—The incalculable number
of petty accidents that concur to produce variability
among brothers, make it impossible to predict the
exact qualities of any individual from hereditary data.
But we may predict average results with great certainty, as will be seen further on, and we can also

obtain precise information concerning the penumbra of uncertainty that attaches itself to single predictions. It would be premature to speak further of this at present ; what has been said is enough to give a clue to the chief motive of this chapter. Its intention has been to show the large part that is always played by chance in the course of hereditary transmission, and to establish the importance of an intelligent use of the laws of chance and of the statistical methods that are based upon them, in expressing the conditions under which heredity acts.

I may here point out that, as the processes of statistics are themselves processes of intimate blendings, their results are the same, whether the materials had been partially blended or not, before they were statistically taken in hand.

CHAPTER III.

ORGANIC STABILITY.

Incipient Structure.—Filial relation.—Stable Forms.—Subordinate positions of Stability.—Model.—Stability of Sports.—Infertility of mixed Types.—Evolution not by minute steps only.

Incipient Structure.—The total heritage of each man must include a greater variety of material than was utilised in forming his personal structure. The existence in some latent form of an unused portion is proved by his power, already alluded to, of transmitting ancestral characters that he did not personally exhibit. Therefore the organised structure of each individual should be viewed as the fulfilment of only one out of an indefinite number of mutually exclusive possibilities. His structure is the coherent and more or less stable development of what is no more than an imperfect sample of a large variety of elements.

The precise conditions under which each several element or particle (whatever may be its nature) finds its way into the sample are, it is needless to repeat, unknown, but we may provisionally classify them under one or other of the following three categories, as they apparently exhaust all reasonable possibilities : first, that in which each element selects its most suitable immediate neighbourhood, in accordance with the guiding idea in Darwin's theory of Pangenesis ; secondly, that of more or less general co-ordination of the influences exerted on each element, not only by its immediate neighbours, but by many or most of the others as well ; finally, that of accident or chance, under which name a group of agencies are to be comprehended, diverse in character and alike only in the fact that their influence on the settlement of each particle was not immediately directed towards that end. In philosophical language we say that such agencies are not purposive, or that they are not teleological ; in popular language they are called accidents or chances.

Filial Relation.—A conviction that inheritance is mainly particulate and much influenced by chance, greatly affects our idea of kinship and makes us consider the parental and filial relation to be curiously circuitous. It appears that there is no direct hereditary relation between the personal parents and the personal child, except perhaps through little-known channels of secondary importance, but that the main line of hereditary connection unites the sets of elements out of which the personal parents had been evolved with the set out of which the personal child was evolved. The main line may be rudely likened to the chain of a necklace, and the personalities to pendants attached to its links. We are unable to see the particles and watch their grouping, and we know nothing directly about them, but we may gain some idea of the various possible results by noting the differences between the brothers in any large fraternity (as will be done further on with much minuteness), whose total heritages must have been much alike, but whose personal structures are often very dissimilar.   This is why it is so important in hereditary inquiry to deal with fraternities rather than with individuals, and with large fraternities rather than small ones. We ought, for example, to compare the group containing both parents and all the uncles and aunts, with that containing all the children. The relative weight to be assigned to the uncles and aunts is a question of detail to be discussed in its proper place further on (see Chap. XI.)

Stable Forms.—The changes in the substance of the newly-fertilised ova of all animals, of which more is annually becoming known,[1] indicate segregations as well as aggregations, and it is reasonable to suppose that repulsions concur with affinities in producing them. We know nothing as yet of the nature of these affinities and repulsions, but we may expect them to act in great numbers and on all sides in a space of three dimensions, just as the personal likings and dis-

---

[1] A valuable memoir on the state of our knowledge of these matters up to the end of 1887 is published in Vol. XIX. of the *Proceedings of the Philosophical Society of Glasgow*, and reprinted under the title of *The Modern Cell Theory, and Theories as to the Physiological Basis of Heredity*, by Prof. John Gray McKendrick, M.D., F.R.S., &c. (R. Anderson, Glasgow, 1888.)

likings of each individual insect in a flying swarm may
be supposed to determine the position that he occupies
in it.   Every particle must have many immediate neighbours.   Even a sphere surrounded by other spheres of
equal sizes, like a cannon-ball in the middle of a heap,
when they are piled in the most compact form, is in
actual contact with no less than twelve others. We may
therefore feel assured that the particles which are still
unfixed must be affected by very numerous influences
acting from all sides and varying with slight changes of
place, and that they may occupy many positions of tem-
porary and unsteady equilibrium, and be subject to
repeated unsettlement, before they finally assume the
positions in which they severally remain at rest.

The whimsical effects of chance in producing stable
results are common enough.   Tangled strings variously
twitched, soon get themselves into tight knots.   Rub-
bish thrown down a sink is pretty sure in time to choke
the pipe ; no one bit may be so large as its bore, but
several bits in their numerous chance encounters will
at length so come into collision as to wedge themselves
into a sort of arch across the tube, and effectually plug
it.   Many years ago there was a fall of large stones from
the ruinous walls of Kenilworth Castle.   Three of them,
if I recollect rightly, or possibly four, fell into a very
peculiar arrangement, and bridged the interval between
the jambs of an old window.   There they stuck fast,
showing clearly against the sky.   The oddity of the
structure attracted continual attention, and its stability
was much commented on.   These hanging stones, as

they were called, remained quite firm for many years ;
at length a storm shook them down.

In every congregation of mutually reacting elements,
some characteristic groupings are usually recognised
that have become familiar through their frequent re-
currence and partial persistence. Being less evanescent
than other combinations, they may be regarded as
temporarily Stable Forms. No demonstration is
needed to show that their number must be greatly
smaller than that of all the possible combinations of
the same elements. I will briefly give as great a
diversity of instances as I can think of, taken from
Governments, Crowds, Landscapes, and even from
Cookery, and shall afterwards draw some illustrations
from Mechanical Inventions, to illustrate what is meant
by characteristic and stable groupings. From some
of them it will also be gathered that secondary and
other orders of stability exist besides the primary
ones.

In Governments, the primary varieties of stable forms
are very few in number, being such as autocracies, con-
stitutional monarchies, oligarchies, or republics. The
secondary forms are far more numerous ; still it is hard
to meet with an instance of one that cannot be pretty
closely paralleled by another. A curious evidence of
the small variety of possible governments is to be found
in the constitutions of the governing bodies of the
Scientific Societies of London and the Provinces, which
are numerous and independent. Their development
seems to follow a single course that has many stages,

and invariably tends to establish the following staff of officers : President, vice-Presidents, a Council, Honorary Secretaries, a paid Secretary, Trustees, and a Treasurer. As Britons are not unfrequently servile to rank, some seek a purely ornamental Patron as well.

Every variety of Crowd has its own characteristic features. At a national pageant, an evening party, a race-course, a marriage, or a funeral, the groupings in each case recur so habitually that it sometimes appears to me as if time had no existence, and that the ceremony in which I am taking part is identical with others at which I had been present one year, ten years, twenty years, or any other time ago.

The frequent combination of the same features in Landscape Scenery, justifies the use of such expressions as "true to nature," when applied to a pictorial composition or to the descriptions of a novel writer. The experiences of travel in one part of the world may curiously resemble those in another. Thus the military expedition by boats up the Nile was planned from experiences gained on the Red River of North America, and was carried out with the aid of Canadian *voyageurs*. The snow mountains all over the world present the same peculiar difficulties to the climber, so that Swiss experiences and in many cases Swiss guides have been used for the exploration of the Himalayas, the Caucasus, the lofty mountains of New Zealand, the Andes, and Greenland. Whenever the general conditions of a new country resemble our own, we recognise characteristic and familiar features at every turn, whether we are walking by the brookside, along the seashore, in the woods, or on the hills.

Even in Cookery it seems difficult to invent a new and good dish, though the current recipes are few, and the proportions of the flour, sugar, butter, eggs, &c., used in making them might be indefinitely varied and be still eatable. I consulted cookery books to learn the facts authoritatively, and found the following passage : "I have constantly kept in view the leading principles of this work, namely, to give in these domestic recipes *the most exact quantities*. . . . I maintain that one cannot be too careful ; it is the only way to put an end to those approximations and doubts which will beset the steps of the inexperienced, and which account for so many people eating indifferent meals at home."[1]

It is the triteness of these experiences that makes the most varied life monotonous after a time, and many old men as well as Solomon have frequent occasion to lament that there is nothing new under the sun.

The object of these diverse illustrations is to impress the meaning I wish to convey, by the phrase of stable forms or groupings, which, however uncertain it may be in outline, is perfectly distinct in substance.

Every one of the meanings that have been attached by writers to the vague but convenient word "type" has for its central idea the existence of a limited number

[1] *The Royal Cookery Book.* By Jules Gouffé, Chef de Cuisine of the Paris Jockey Club ; translated by Alphonse Gouffé, Head Pastry Cook to H.M. the Queen. Sampson Low. 1869. Introduction, p. 9.

of frequently recurrent forms. The word etymologically
compares these forms to the identical medals that may
be struck by one or other of a set of dies. The central
idea on which the phrase " stable forms " is based is of
the same kind, while the phrase further accounts for
their origin, vaguely it may be, but still significantly,
by showing that though we know little or nothing of
details, the result of organic groupings is analogous to
much that we notice elsewhere on every side.

Subordinate positions of Stability.—Of course there
are different degrees of stability. If the same structural
form recurs in successively descending generations, its
stability must be great, otherwise it could not have
withstood the effects of the admixture of equal doses of
alien elements in successive generations. Such a form
well deserves to be called typical. A breeder would
always be able to establish it. It tends of itself to
become a new and stable variety ; therefore all the
breeder has to attend to is to give fair play to its
tendency, by weeding out from among its offspring such
reversions to other forms as may crop up from time to
time, and by preserving the breed from rival admixtures
until it has become confirmed, and adapted in every
minute particular to its surroundings.

Personal Forms may be compared to Human Inventions, as these also may be divided into types, sub-types,
and deviations from them. Every important invention is a new type, and of such a definite kind as to
admit of clear verbal description, and so of becoming

the subject of patent rights ; at the same time it need
not be so minutely defined as to exclude the possibility
of small improvements or of deviations from the main
design, any of which may be freely adopted by the inventor without losing the protection of his patent.   But
the range of protection is by no means sharply distinct,
as most inventors know to their cost.   Some other man,
who may or may not be a plagiarist, applies for a separate patent for himself, on the ground that he has introduced modifications of a fundamental character ; in other
words, that he has created a fresh type.   His application
is opposed, and the question whether his plea be valid
or not, becomes a subject for legal decision.

Whenever a patent is granted subsidiary to another,
and lawful to be used only by those who have acquired
rights to work the primary invention, then we should
rank the new patent as a secondary and not as a
primary type.   Thus we see that mechanical inventions
offer good examples of types, sub-types, and mere
deviations.

The three kinds of public carriages that characterise
the streets of London ; namely, omnibuses, hansoms,
and four-wheelers, are specific and excellent illustrations of what I wish to express by mechanical types,
as distinguished from sub-types.   Attempted improvements in each of them are yearly seen, but none have as
yet superseded the old familiar patterns, which cannot,
as it thus far appears, be changed with advantage, taking
the circumstances of London as they are.   Yet there
have been numerous subsidiary and patented contriv-

ances, each a distinct step in the improvement of one
or other of the three primary types, and there are or
may be in each of the three an indefinite number
of varieties in details, too unimportant to be subjects
of patent rights.

The broad classes, of primary or subordinate types,
and of mere deviations from them, are separated by no
well-defined frontiers. Still the distinction is very serviceable, so much so that the whole of the laws of patent
and copyright depend upon it, and it forms the only
foundation for the title to a vast amount of valuable
property. Corresponding forms of classification must
be equally appropriate to the organic structure of all
living things.

Model.—The distinction between primary and subordinate positions of stability will be made clearer by the

FIG. I.

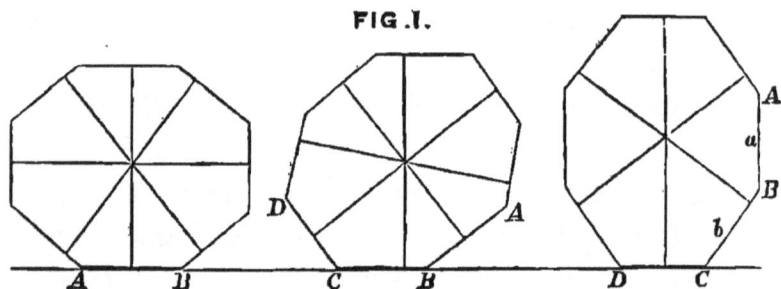

help of Fig 1, which is drawn from a model I made. The
model has more sides, but Fig. 1 suffices for illustration.
It is a polygonal slab that can be made to stand on any
one of its edges when set upon a level table, and is

intended to illustrate the meaning of primary and sub-ordinate stability in organic structures, although the conditions of these must be far more complex than anything we have wits to imagine. The model and the organic structure have the cardinal fact in common, that if either is disturbed without transgressing the range of its stability, it will tend to re-establish itself, but if the range is overpassed it will topple over into a new position ; also that both of them are more likely to topple over towards the position of primary stability, than away from it.

The ultimate point to be illustrated is this. Though a long established race habitually breeds true to its kind, subject to small unstable deviations, yet every now and then the offspring of these deviations do not tend to revert, but possess some small stability of their own. They therefore have the character of sub-types, always, however, with a reserved tendency under strained conditions, to revert to the earlier type. The model further illustrates the fact that sometimes a sport may occur of such marked peculiarity and stability as to rank as a new type, capable of becoming the origin of a new race with very little assistance on the part of natural selection. Also, that a new type may be reached without any large single stride, but through a fortunate and rapid succession of many small ones.

The model is a polygonal slab, the polygon being one that might have been described within an oval, and it is so shaped as to stand on any one of its edges. When the slab rests as in Fig. 1, on the edge A B, corresponding to

the shorter diameter of the oval, it stands in its most
stable position, and in one from which it is equally difficult to dislodge it by a tilt either forwards or backwards.
So long as it is merely tilted it will fall back on being
left alone, and its position when merely tilted corresponds to a simple deviation. But when it is pushed
with sufficient force, it will tumble on to the next
edge, B C, into a new position of stability. It will
rest there, but less securely than in its first position;
moreover its range of stability will no longer be disposed symmetrically. A comparatively slight push from
the front will suffice to make it tumble back, a comparatively heavy push from behind is needed to make
it tumble forward. If it be tumbled over into a
third position (not shown in the Fig.), the process
just described may recur with exaggerated effect, and
similarly for many subsequent ones. If, however, the
slab is at length brought to rest on the edge C D,
most nearly corresponding to its longest diameter, the
next onward push, which may be very slight, will suffice
to topple it over into an entirely new system of stability;
in other words, a "sport" comes suddenly into existence. Or the figure might have been drawn with its
longest diameter passing into a projecting spur, so that
a push of extreme strength would be required to topple
it entirely over.

If the first position, A B, is taken to represent a type,
the other portions will represent sub-types. All the
stable positions on the same side of the longer diameter
are subordinate to the first position. On whichever of

of them the polygon may stand, its principal tendency on being seriously disturbed will be to fall back towards the first position ; yet each position is stable within certain limits.

Consequently the model illustrates how the following conditions may co-exist : (1) Variability within narrow limits without prejudice to the purity of the breed. (2) Partly stable sub-types. (3) Tendency, when much disturbed, to revert from a sub-type to an earlier form. (4) Occasional sports which may give rise to new types.

Stability of Sports.—Experience does not show that those wide varieties which are called " sports " are unstable. On the contrary, they are often transmitted to successive generations with curious persistence. Neither is there any reason for expecting otherwise. While we can well understand that a strained modification of a type would not be so stable as one that approximates more nearly to the typical centre, the variety may be so wide that it falls into different conditions of stability, and ceases to be a strained modification of the original type.

The hansom cab was originally a marvellous novelty. In the language of breeders it was a sudden and remarkable " sport," yet the suddenness of its appearance has been no bar to its unchanging hold on popular favour. It is not a monstrous anomaly of incongruous parts, and therefore unstable, but quite the contrary. Many other instances of very novel and yet stable inventions could be quoted. One of the earliest

electrical batteries was that which is still known as a
Grove battery, being the invention of Sir William Grove.
Its principle was quite new at the time, and it continues
in use without alteration.

The persistence in inheritance of trifling characteristics,
such as a mole, a white tuft of hair, or multiple fingers,
has often been remarked. The reason of it is, I presume,
that such characteristics have inconsiderable influence
upon the general organic stability; they are mere
excrescences, that·may be associated with very different
types, and are therefore inheritable without let or
hindrance.

It seems to me that stability of type, about which we
as yet know very little, must be an important factor in
the general theory of heredity, when the theory is
applied to cases of high breeding. It will be shown
later on, at what point a separate allowance requires
to be made for it. But in the earlier and principal
part of the inquiry, which deals with the inheritance of
qualities that are only exceptional in a small degree, a
separate allowance does not appear to be required.

Infertility of Mixed Types.—It is not difficult to see
in a general way why very different types should refuse
to coalesce, and it is scarcely possible to explain the
reason why, more clearly than by an illustration. Thus
a useful blend between a four-wheeler and a hansom
would be impossible; it would have to run on three
wheels and the half-way position for the driver would
be upon its roof. A blend would be equally impossible

between an omnibus and a hansom, and it would be
difficult between an omnibus and a four-wheeler.

Evolution not by Minute Steps Only.—The theory
of Natural Selection might dispense with a restriction,
for which it is difficult to see either the need or the
justification, namely, that the course of evolution always
proceeds by steps that are severally minute, and that
become effective only through accumulation. That
the steps *may* be small and that they *must* be small are
very different views; it is only to the latter that I
object, and only when the indefinite word " small " is used
in the sense of " barely discernible," or as small compared with such large sports as are known to have been
the origins of new races. An apparent ground for the
common belief is founded on the fact that whenever
search is made for intermediate forms between widely
divergent varieties, whether they be of plants or of
animals, of weapons or utensils, of customs, religion or
language, or of any other product of evolution, a long
and orderly series can usually be made out, each member
of which differs in an almost imperceptible degree from
the adjacent specimens. But it does not at all follow
because these intermediate forms have been found to
exist, that they are the very stages that were passed
through in the course of evolution. Counter evidence
exists in abundance, not only of the appearance of considerable sports, but of their remarkable stability in
hereditary transmission. Many of the specimens of
intermediate forms may have been unstable varieties,

whose descendants had reverted; they might be looked
upon as tentative and faltering steps taken along parallel
courses of evolution, and afterwards retraced. Affiliation
from each generation to the next requires to be proved
before any apparent line of descent can be accepted
as the true one. The history of inventions fully illustrates this view. It is a most common experience that
what an inventor knew to be original, and believed to
be new, had been invented independently by others
many times before, but had never become established.
Even when it has new features, the inventor usually
finds, on consulting lists of patents, that other inventions
closely border on his own. Yet we know that inventors
often proceed by strides, their ideas originating in some
sudden happy thought suggested by a chance occurrence,
though their crude ideas may have to be laboriously
worked out afterwards. If, however, all the varieties of
any machine that had ever been invented, were collected
and arranged in a Museum in the apparent order of
their Evolution, each would differ so little from its
neighbour as to suggest the fallacious inference that the
successive inventors of that machine had progressed by
means of a very large number of hardly discernible
steps.

The object of this and of the preceding chapter has
been first to dwell on the fact of inheritance being
"particulate," secondly to show how this fact is com-
patible with the existence of various types, some of
which are subordinate to others, and thirdly to argue

D

that Evolution need not proceed by small steps only. I have largely used metaphor and illustration to explain the facts, wishing to avoid entanglements with theory as far as possible, inasmuch as no complete theory of inheritance has yet been propounded that meets with general acceptation.

CHAPTER IV.

SCHEMES OF DISTRIBUTION AND OF FREQUENCY.

Fraternities and Populations to be treated as Units.—Schemes of Distribution and their Grades.—The Shape of Schemes is independent of the number of observations.—Data for Eighteen Schemes.—Application of the method of Schemes to inexact Measures.—Schemes of Frequency.

Fraternities and Populations to be Treated as Units.— The science of heredity is concerned with Fraternities and large Populations rather than with individuals, and must treat them as units. A compendious method is therefore requisite by which we may express the distribution of each faculty among the members of any large group, whether it be a Fraternity or an entire Population.

The knowledge of an average value is a meagre piece of information. How little is conveyed by the bald statement that the average income of English families is 100*l.* a year, compared with what we should learn if we were told how English incomes were distributed ; what proportion of our countrymen had just and only just enough means to ward off starvation, and what were the

proportions of those who had incomes in each and every
other degree, up to the huge annual receipts of a few
great speculators, manufacturers, and landed proprietors.
So in respect to the distribution of any human quality
or faculty, a knowledge of mere averages tells but little;
we want to learn how the quality is distributed among
the various members of the Fraternity or of the Population, and to express what we know in so compact a
form that it can be easily grasped and dealt with.
A parade of great accuracy is foolish, because precision
is unattainable in biological and social statistics; their
results being never strictly constant. Over-minuteness
is mischievous, because it overwhelms the mind with
more details than can be compressed into a single
view. We require no more than a fairly just and
comprehensive method of expressing the way in which
each measurable quality is distributed among the
members of any group, whether the group consists
of brothers or of members of any particular social,
local, or other body of persons, or whether it is co-
extensive with an entire nation or race.

A knowledge of the distribution of any quality enables us to ascertain the Rank that each man holds
among his fellows, in respect to that quality. This is
a valuable piece of knowledge in this struggling and
competitive world, where success is to the foremost, and
failure to the hindmost, irrespective of absolute efficiency.
A blurred vision would be above all price to an individual man in a nation of blind men, though it would
hardly enable him to earn his bread elsewhere. When

the distribution of any faculty has been ascertained, we can tell from the measurement, say of our child, how he ranks among other children in respect to that faculty, whether it be a physical gift, or one of health, or of intellect, or of morals. As the years go by, we may learn by the same means whether he is making his way towards the front, whether he just holds his place, or whether he is falling back towards the rear. Similarly as regards the position of our class, or of our nation, among other classes and other nations.

Schemes of Distribution and their Grades.—I shall best explain my graphical method of expressing Distribution, which I like the more, the more I use it, and which I have latterly much developed, by showing how to determine the Grade of an individual among his fellows in respect to any particular faculty. Suppose that we have already put on record the measures of many men in respect to Strength, exerted as by an archer in pulling his bow, and tested by one of Salter's well-known dial instruments with a movable index. Some men will have been found strong and others weak ; how can we picture in a compendious diagram, or how can we define by figures, the distribution of this faculty of Strength throughout the group? How shall we determine and specify the Grade that any particular person would occupy in the group? The first step is to marshal our measures in the orderly way familiar to statisticians, which is shown in Table I. I usually work to about twice its degree of minuteness, but enough has been entered in the Table for the purpose of illustration, while its small size makes it all the more intelligible.

The fourth column of the Table headed " Percentages " of " Sums from the beginning," is pictorially translated into Fig. 2, and the third column headed " Percentages " of " No. of cases observed," into Fig. 3. The scale of

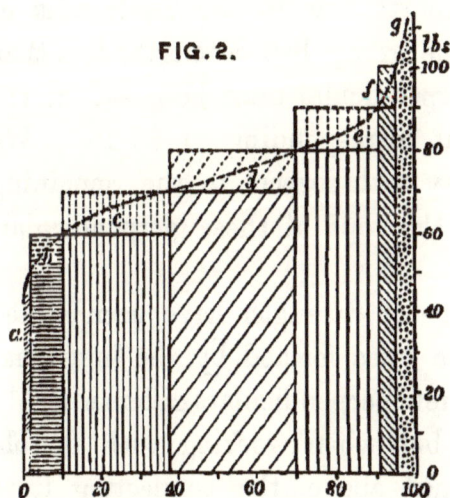

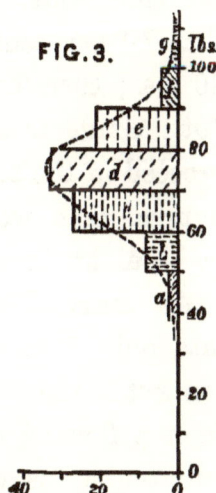

lbs. is given at the side of both Figs. : and the compartments *a* to *g*, that are shaded with *broken* lines, have the same meaning in both, but they are differently disposed in the two Figs. We will now consider Fig. 2 only, which is the one that principally concerns us. The percentages in the last column of Table I. have been marked off on the bottom line of Fig. 2, where they are called (centesimal) Grades. The number of lbs. found in the first column of the Table determines the height of the vertical lines to be erected at the corresponding Grades when we are engaged in constructing the Figure.

Let us begin with the third line in the Table for illustration: it tells us that 37 per cent. of the group had Strengths less than 70 lbs. Therefore, when drawing the figure, a perpendicular must be raised at the 37th grade to a height corresponding to that of 70 lbs. on the side scale. The fourth line in the Table tells us that 70 per cent. of the group had Strengths less than 80 lbs. ; therefore a perpendicular must be raised at the 70th Grade to a height corresponding to 80 lbs. We proceed in the same way with respect to the remaining figures, then we join the tops of these perpendiculars by straight lines.

As these observations of Strength have been sorted into only 7 groups, the trace formed by the lines that connect the tops of the few perpendiculars differs sensibly from a flowing curve, but when working with double minuteness, as mentioned above, the connecting lines differ little to the eye from the dotted curve. The dotted curve may then be accepted as that which would result if a separate perpendicular had been drawn for every observation, and if permission had been given to slightly smooth their irregularities. I call the figure that is bounded by such a curve as this, a Scheme of Distribution; the perpendiculars that formed the scaffolding by which it was constructed having been first rubbed out. (See Fig 4, next page.)

A Scheme enables us in a moment to find the Grade

of Rank (on a scale reckoned from 0° to 100°) of any
person in the group to which he belongs. The measured
strength of the person is to be looked for in the side
scale of the Scheme; a horizontal line is thence drawn
until it meets the curve; from the point of meeting
a perpendicular is dropped upon the scale of Grades
at the base; then the Grade on which it falls is

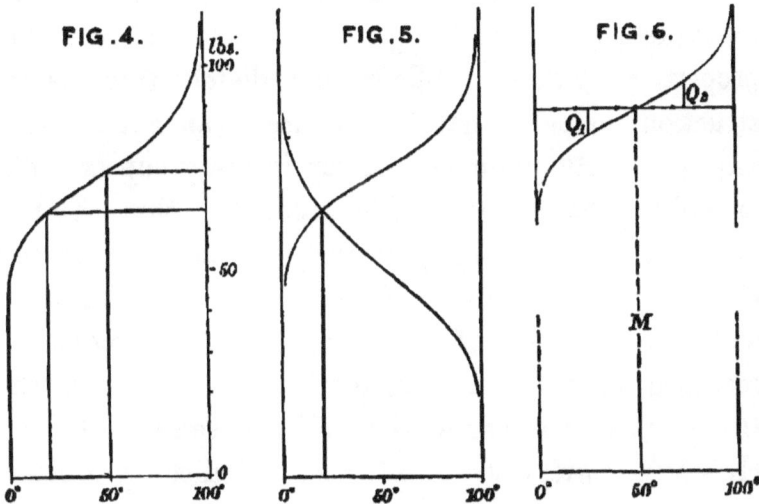

the one required. For example: let us suppose the
Strength of Pull of a man to have been 74 lbs.,
and that we wish to determine his Rank in Strength
among the large group of men who were measured
at the Health Exhibition in 1884. We find by Fig.
4 that his centesimal Grade is 50°; in other words,
that 50 per cent. of the group will be weaker than
he is, and 50 per cent. will be stronger. His

position will be exactly Middlemost, after the Strengths of all the men in the group have been marshalled in the order of their magnitudes. In other words, he is of mediocre strength. The accepted term to express the value that occupies the Middlemost position is "Median," which may be used either as an adjective or as a substantive, but it will be usually replaced in this book by the abbreviated form M. I also use the word "Mid" in a few combinations, such as "Mid-Fraternity," to express the same thing. The Median, M, has three properties. The first follows immediately from its construction, namely, that the chance is an equal one, of any previously unknown measure in the group exceeding or falling short of M. The second is, that the most probable value of any previously unknown measure in the group is M. Thus if N be any one of the measures, and u be the value of the unit in which the measure is recorded, such as an inch, tenth of an inch, &c., then the number of measures that fall between $(N - \frac{1}{2}u)$ and $(N + \frac{1}{2}u)$, is greatest when $N = M$. Mediocrity is always the commonest condition, for reasons that will become apparent later on. The third property is that whenever the curve of the Scheme is symmetrically disposed on either side of M, except that one half of it is turned upwards, and the other half downwards, then M is identical with the ordinary Arithmetic Mean or Average. This is closely the condition of all the curves I have to discuss. The reader may look on the Median and on the Mean as being practically the same things, throughout this book.

It must be understood that M, like the Mean or the
Average, is almost always an interpolated value, corresponding to no real measure.   If the observations were
infinitely numerous its position would not differ more
than infinitesimally from that of some one of them;
even in a series of one or two hundred in number, the
difference is insignificant.

Now let us make our Scheme answer another question.
Suppose we want to know the percentage of men in the
group of which we have been speaking, whose Strength
lies between any two specified limits, as between 74 lbs.
and 64 lbs.   We draw horizontal lines (Fig. 4) from
points on the side scale corresponding to either limit,
and drop perpendiculars upon the base, from the points
where those lines meet the curve.   Then the number of
Grades in the intercept is the answer.   The Fig. shows
that the number in the present case is 30 ; therefore
30 per cent. of the group have Strengths of Pull ranging
between 74 and 64 lbs.

We learn how to transmute female measures of any
characteristic into male ones, by comparing their respective schemes, and devising a formula that will change
the one into the other.   In the case of Stature, the
simple multiple of 1·08 was found to do this with
sufficient precision.

If we wish to compare the average Strengths of two
different groups of persons, say one consisting of men
and the other of women, we have simply to compare
the values at the 50th Grades in the two schemes.   For
even if the Medians differ considerably from the Means,

both the ratios and the differences between either pair of values would be sensibly the same.

A different way of comparing two Schemes is sometimes useful. It is to draw them in opposed directions, as in Fig. 5, p. 40. Their curves will then cut each other at some point, whose Grade when referred to either of the two Schemes (whichever of them may be preferred), determines the point at which the same values are to be found. In Fig. 5, the Grade in the one Scheme is 20°; therefore in the other Scheme it is 100°−20°, or 80°. In respect to the Strength of Pull of men and women, it appears that the woman who occupies the Grade of 96° in her Scheme, has the same strength as the man who occupies the Grade of 4° in his Scheme.

I should add that this great inequality in Strength between the sexes, is confirmed by other measurements made at the same time in respect to the Strength of their Squeeze, as tested by another of Salter's instruments. Then the woman in the 93rd and the man in the 7th Grade of their resective Schemes, proved to be of equal strength. In my paper[1] on the results obtained at the laboratory, I remarked: "Very powerful women exist, but happily perhaps for the repose of the other sex such gifted women are rare. Out of 1,657 adult women of all ages measured at the laboratory, the strongest could only exert a squeeze of 86 lbs., or about that of a medium man."

[1] *Journ. Anthropol. Inst.* 1885. *Mem.:* There is a blunder in the paragraph, p. 23, headed "Height Sitting and Standing." The paragraph should be struck out.

*The Shape of Schemes is Independent of the Number
of Observations.*—When Schemes are drawn from different samples of the same large group of measurements,
though the number in the several samples may differ
greatly, we can always so adjust the horizontal scales
that the breadth of the several Schemes shall be uniform.
Then the shapes of the Schemes drawn from different
samples will be little affected by the number of observations used in each, supposing of course that the numbers
are never too small for ordinary statistical purposes.
The only recognisable differences between the Schemes
will be, that, if the number of observations in the
sample is very large, the upper margin of the Scheme
will fall into a more regular curve, especially towards
either of its limits.   Some irregularity will be found in
the above curve of the Strength of Pull; but if the
observations had been ten times more numerous, it is
probable, judging from much experience of such curves,
that the irregularity would have been less conspicuous,
and perhaps would have disappeared altogether.

However numerous the observations may be, the
curve will always be uncertain and incomplete at its
extreme ends, because the next value may happen to be
greater or less than any one of those that preceded it.
Again, the position of the first and the last observation,
supposing each observation to have been laid down separately, can never coincide with the adjacent limit.   The
more numerous the observations, and therefore the closer
the perpendiculars by which they are represented, the
nearer will the two extreme perpendiculars approach the

limits, but they will never actually touch them. A chess
board has eight squares in a row, and eight pieces may be
arranged in order on any one row, each piece occupying
the centre of a square. Let the divisions in the row be
graduated, calling the boundary to the extreme left,
0°. Then the successive divisions between the squares
will be 1°, 2°, 3°, up to 7°, and the boundary to the
extreme right will be 8°. It is clear that the position of
the first piece lies half-way between the grades (in a
scale of eight grades) of 0° and 1°; therefore the grade
occupied by the first piece would be counted on that
scale as 0·5°; also the grade of the last piece as 7·5°.
Or again, if we had 800 pieces, and the same number
of class-places, the grade of the first piece, in a scale
of 800 grades, would exceed the grade 0°, by an amount
equal to the width of one half-place on that scale,
while the last of them would fall short of the 800th
grade by an equal amount. This half-place has to be
attended to and allowed for when schemes are constructed from comparatively few observations, and
always when values that are very near to either of the
centesimal grades 0° or 100° are under observation ;
but between the centesimal grades of 5° and 95° the
influence of a half class-place upon the value of the
corresponding observation is insignificant, and may be
disregarded. It will not henceforth be necessary to
repeat the word centesimal. It will be always implied
when nothing is said to the contrary, and nothing
henceforth will be said to the contrary. The word will
be used for the last time in the next paragraph.

Data for Eighteen Schemes.—Sufficient data for re-constructing any Scheme, with much correctness, may be printed in a single line of a Table, and according to a uniform plan that is suitable for any kind of values. The measures to be recorded are those at a few definite Grades, beginning say at 5°, ending at 95°, and including every intermediate tenth Grade from 10° to 90°. It is convenient to add those at the Grades 25° and 75°, if space permits. The former values are given for eighteen different Schemes, in Table 2. In the memoir from which that table is reprinted, the values at what I now call (centesimal) Grades, were termed Percentiles. Thus the values at the Grades 5° and 10° would be respectively the 5th and the 10th percentile. It still seems to me that the word percentile is a useful and expressive abbreviation, but it will not be necessary to employ it in the present book. It is of course unadvisable to use more technical words than is absolutely necessary, and it will be possible to get on without it, by the help of the new and more important word " Grade."

A series of Schemes that express the distribution of various faculties, is valuable in an anthropometric labora-tory, for they enable every person who is measured to find his Rank or Grade in each of them.

Diagrams may also be constructed by drawing parallel lines, each divided into 100 Grades, and entering each round number of inches, lbs., &c., at their proper places. A diagram of this kind is very convenient for reference, but it does not admit of being printed; it must be drawn or lithographed. I have constructed one of these

from the 18 Schemes, and find it is easily understood
and much used at my laboratory.

Application of Schemes to Inexact Measures.—Schemes
of Distribution may be constructed from observations
that are barely exact enough to deserve to be called
measures.

I will illustrate the method of doing so by marshalling
the data contained in a singularly interesting little
memoir written by Sir James Paget, into the form of
such a Scheme. The memoir is published in vol. v. of St.
Bartholomew's Hospital Reports, and is entitled " What
Becomes of Medical Students." He traced with great
painstaking the career of no less than 1,000 pupils who
had attended his classes at that Hospital during various
periods and up to a date 15 years previous to that at
which his memoir was written. He thus did for St.
Bartholomew's Hospital what has never yet been done,
so far as I am aware, for any University or Public
School, whose historians count the successes and are
silent as to the failures, giving to inquirers no adequate
data for ascertaining the real value of those institutions
in English Education. Sir J. Paget divides the successes
of his pupils in their profession into five grades, all of
which he carefully defines; they are *distinguished;
considerable; moderate; very limited success;* and
failures. Several of the students had left the profession either before or after taking their degrees, usually
owing to their unfitness to succeed, so after analysing
the accounts of them given in the memoir, I drafted

several into the list of failures and distributed the rest, with the result that the number of cases in the successive classes, amounting now to the full total of 1,000, became 28, 80, 616, 151, and 125. This differs, I should say, a little from the inferences of the author, but the matter is here of small importance, so I need not go further into details.

If a Scheme is drawn from these figures, in the way described in page 39, it will be found to have the characteristic shape of our familiar curve of Distribution. If we wished to convey the utmost information that this Scheme is capable of giving, we might record in much detail the career of two or three of the men who are clustered about each of a few selected Grades, such as those that are used in Table II., or fewer of them. I adopted this method when estimating the variability of the Visualising Power (*Inquiries into Human Faculty*). My data were very lax, but this method of treatment got all the good out of them that they possessed. In the present case, it appears that towards the foremost of the successful men within fifteen years of taking their degrees, stood the three Professors of Anatomy at Oxford, Cambridge, and Edinburgh; that towards the bottom of the failures, lay two men who committed suicide under circumstances of great disgrace, and lowest of all Palmer, the Rugeley murderer, who was hanged.

We are able to compare any two such Schemes as the above, with numerical precision. The want of exactness in the data from which they are drawn, will of course cling to the result, but no new error will be introduced

by the process of comparison. Suppose the second
Scheme to refer to the successes of students from another
hospital, we should draw the two Schemes in opposed
directions, just as was done in the Strength of Pull of
Males and Females, Fig. 5, and determine the Grade
in either of the Schemes at which success was equal.

Schemes of Frequency.—The method of arranging
observations in an orderly manner that is generally
employed by statisticians, is shown in Fig. 3, page 38,
which expresses the same facts as Fig. 2 under a different
aspect, and so gives rise to the well-known Curve of
" Frequency of Error," though in Fig. 3 the curve is
turned at right angles to the position in which it is
usually drawn. It is so placed in order to show more
clearly its relation to the Curve of Distribution. The
Curve of Frequency is far less convenient than that of
Distribution, for the purposes just described and for
most of those to be hereafter spoken of. But the Curve
of Frequency has other uses, of which advantage will
be taken later on, and to which it is unnecessary now
to refer.

A Scheme as explained thus far, is nothing more than
a compendium of a mass of observations which, on being
marshalled in an orderly manner, fall into a diagram
whose contour is so regular, simple, and bold, as to
admit of being described by a few numerals (Table 2),
from which it can at any time be drawn afresh. The
regular distribution of the several faculties among a
large population is little disturbed by the fact that its

members are varieties of different types and sub-types. So the distribution of a heavy mass of foliage gives little indication of its growth from separate twigs, of separate branches, of separate trees.

The application of theory to Schemes, their approximate description by only two values, and the properties of their bounding Curves, will be described in the next chapter.

CHAPTER V.

NORMAL VARIABILITY.

Schemes of Deviations.—Normal Curve of Distribution.—Comparison of the observed with the Normal Curve.—The value of a single Deviation at a known Grade determines a Normal Scheme of Deviations.— Two Measures at two known Grades determine a Normal Scheme of Measures.—The Charms of Statistics.—Mechanical illustration of the Cause of the Curve of Frequency.—Order in apparent Chaos.— Problems in the Law of Error.

Schemes of Deviations.—We have now seen how easy it is to represent the distribution of any quality among a multitude of men, either by a simple diagram or by a line containing a few figures. In this chapter it will be shown that a considerably briefer description is approximately sufficient.

Every measure in a Scheme is equal to its Middlemost, or Median value, or M, *plus* or *minus* a certain Deviation from M. The Deviation, or "Error" as it is technically called, is *plus* for all grades above 50°, zero for 50°, and *minus* for all grades below 50°. Thus if (±D) be the deviation from M in any particular case, every measure in a Scheme may be expressed in the

E 2

form of $M + (\pm D)$. If $M = O$, or if it is subtracted from every measure, the residues which are the different values of $(\pm D)$ will form a Scheme by themselves. Schemes may therefore be made of Deviations as well as of Measures, and one of the former is seen in the upper part of Fig. 6, page 40. It is merely the upper portion of the corresponding Scheme of Measures, in which the axis of the curve plays the part of the base.

A strong family likeness runs between the 18 different Schemes of Deviations that may be respectively derived from the data in the 18 lines of Table 2. If the slope of the curve in one Scheme is steeper than that of another, we need only to fore-shorten the steeper Scheme, by inclining it away from the line of sight, in order to reduce its apparent steepness and to make it look almost identical with the other. Or, better still, we may select appropriate vertical scales that will enable all the Schemes to be drawn afresh with a uniform slope, and be made strictly comparable.

Suppose that we have only two Schemes, A. and B., that we wish to compare. Let $L._1$, $L._2$ be the lengths of the perpendiculars at two specified grades in Scheme A., and $K._1$ $K._2$ the lengths of those at the same grades in Scheme B.; then if every one of the data from which Scheme B. was drawn be multiplied by $\dfrac{L._1 - L._2}{K._1 - K._2}$, a series of transmuted data will be obtained for drawing a new Scheme B′., on such a vertical scale that its general slope between the selected grades shall be the same as in Scheme A. For practical convenience the selected Grades will be always those of 25° and 75°. They stand at the first and third quarterly divisions of the base, and are therefore easily found by a pair of compasses. They are also well placed to afford a fair criterion of the general slope of the Curve. If we call the perpendicular at 25°, $Q_{.1}$; and that at 75°, $Q_{.2}$, then the unit by which every Scheme will be defined is its value of $\frac{1}{2}(Q_{.2} - Q_{.1})$, and will be called its Q. As the M measures the Average Height of the curved boundary of a Scheme, so the Q measures its general slope. When we wish to transform many different Schemes, numbered I., II., III., &c., whose respective values of Q are q_1, q_2, q_3, &c., to others whose values of Q are in each case equal to q_0, then all the data from which

Scheme I. was drawn, must be multiplied by $\frac{q_0}{q_1}$; those

from which Scheme II. was drawn, by $\frac{q_0}{q_2}$, and so on, and

new Schemes have to be constructed from these transmuted values.

Our Q has the further merit of being practically the same as the value which mathematicians call the "Probable Error," of which we shall speak further on.

Want of space in Table 2 prevented the insertion of the measures at the Grades 25° and 75°, but those at 20° and 30° are given on the one hand, and those at 70° and 80° on the other, whose respective averages differ but little from the values at 25° and 75°. I therefore will use those four measures to obtain a value for our unit, which we will call Q', to distinguish it from Q.

These are not identical in value, because the outline of
the Scheme is a curved and not a straight line, but the
difference between them is small, and is approximately
the same in all Schemes. It will shortly be seen that
$Q' = 1·015 \times Q$ approximately; therefore a series of Deviations measured in terms of the large unit Q' are
numerically smaller than if they had been measured in
terms of the small unit (for the same reason that the
numerals in 2, 3, &c., *feet* are smaller than those in the
corresponding values of 24, 36, &c., *inches*), and they
must be multiplied by 1.015 when it is desired to
change them into a series having the smaller value of Q
for their unit.

All the 18 Schemes of Deviation that can be derived
from Table 2 have been treated on these principles, and
the results are given in Table 3. Their general accordance with one another, and still more with the mean of
all of them, is obvious.

Normal Curve of Distribution.—The values in the
bottom line of Table 3, which is headed " Normal Values
when $Q = 1$," and which correspond with minute precision to those in the line immediately above them, are
not derived from observations at all, but from the well-
known Tables of the " Probability Integral " in a way
that mathematicians will easily understand by comparing
the Tables 4 to 8 inclusive. I need hardly remind the
reader that the Law of Error upon which these Normal
Values are based, was excogitated for the use of astro-
nomers and others who are concerned with extreme

accuracy of measurement, and without the slightest idea
until the time of Quetelet that they might be applicable
to human measures.   But Errors, Differences, Deviations,
Divergencies, Dispersions, and individual Variations, all
spring from the same kind of causes.   Objects that bear
the same name, or can be described by the same phrase,
are thereby acknowledged to have common points of
resemblance, and to rank as members of the same species,
class, or whatever else we may please to call the group.
On the other hand, every object has Differences peculiar
to itself, by which it is distinguished from others.

This general statement is applicable to thousands of
instances.   The Law of Error finds a footing wherever
the individual peculiarities are wholly due to the combined influence of a multitude of "accidents," in the
sense in which that word has already been defined.
All persons conversant with statistics are aware that
this supposition brings Variability within the grasp
of the laws of Chance, with the result that the
relative frequency of Deviations of different amounts
admits of being calculated, when those amounts are
measured in terms of any self-contained unit of variability, such as our Q.   The Tables 4 to 8 give the
results of these purely mathematical calculations, and
the Curves based upon them may with propriety be
distinguished as "Normal."   Tables 7 and 8 are based
upon the familiar Table of the Probability Integral,
given in Table 5, *vid* that in Table 6, in which the unit
of variability is taken to be the "Probable Error" or
our Q, and not the "Modulus."   Then I turn Table 6

inside out, as it were, deriving the "arguments" for
Tables 7 and 8 from the entries in the body of Table 6,
and making other easily intelligible alterations.

Comparison of the Observed with the Normal Curve.
—I confess to having been amazed at the extraordinary
coincidence between the two bottom lines of Table 3,
considering the great variety of faculties contained in
the 18 Schemes; namely, *three kinds of linear measurement*, besides one of weight, one of capacity, two of
strength, one of vision, and one of swiftness. It is
obvious that weight cannot really vary at the same rate
as height, even allowing for the fact that tall men are
often lanky, but the theoretical impossibility is of the
less practical importance, as the variations in weight are
small compared to the weight itself. Thus we see from
the value of Q in the first column of Table 3, that half
of the persons deviated from their M by no more than
10 or 11 lbs., which is about one-twelfth part of the
value of M. Although the several series in Table 3 run
fairly well together, I should not have dared to hope
that their irregularities would have balanced one another
so beautifully as they have done. It has been objected
to some of my former work, especially in *Hereditary
Genius*, that I pushed the applications of the Law of
Frequency of Error somewhat too far. I may have done
so, rather by incautious phrases than in reality; but
I am sure that, with the evidence now before us, the
applicability of that law is more than justified within
the reasonable limits asked for in the present book. I

am satisfied to claim that the Normal Curve is a fair
average representation of the Observed Curves during
nine-tenths of their course; that is, for so much of
them as lies between the grades of 5° and 95°.   In
particular, the agreement of the Curve of Stature with
the Normal Curve is very fair, and forms a mainstay of
my inquiry into the laws of Natural Inheritance.

It has already been said that mathematicians laboured
at the law of Error for one set of purposes, and we
are entering into the fruits of their labours for another.
Hence there is no ground for surprise that their Nomenclature is often cumbrous and out of place, when applied
to problems in heredity.   This is especially the case
with regard to their term of " Probable Error," by which
they mean the value that one half of the Errors exceed
and the other half fall short of.   This is practically the
same as our Q.[1]   It is strictly the same whenever the
two halves of the Scheme of Deviations to which it
applies are symmetrically disposed about their common
axis.

The term Probable Error, in its plain English interpretation of the *most* Probable Error, is quite misleading, for it is *not* that.   The most Probable Error
(as Dr. Venn has pointed out, in his *Logic of Chance*)

---

[1] The following little Table may be of service :—

*Values of the different Constants when the Prob. Error is taken as unity, and
their corresponding Grades.*

Prob. Error .... 1·000 ; corresponding Grades 25°·0, 75°·0
Modulus .... 2·097 ;   „   „   7°·9, 92°·1
Mean Error............... 1·183 ;   „   „   21°·2, 78°·8
Error of Mean Squares 1·483 ;   „   „   16°·0, 84°·0

is zero. This results from what was said a few pages back about the most probable measure in a Scheme being its M. In a Scheme of Errors the M is equal to 0, therefore the most Probable Error in such a Scheme is 0 also. It is astonishing that mathematicians, who are the most precise and perspicacious of men, have not long since revolted against this cumbrous, slip-shod, and misleading phrase. They really mean what I should call the Mid-Error, but their phrase is too firmly established for me to uproot it. I shall however always write the word Probable when used in this sense, in the form of "Prob."; thus "Prob. Error," as a continual protest against its illegitimate use, and as some slight safeguard against its misinterpretation. Moreover the term Probable Error is absurd when applied to the subjects now in hand, such as Stature, Eye-colour, Artistic Faculty, or Disease. I shall therefore usually speak of Prob. Deviation.

Though the value of our Q is the same as that of the Prob. Deviation, Q is not a convertible term with Prob. Deviation. We shall often have to speak of the one without immediate reference to the other, just as we speak of the diameter of the circle without reference to any of its properties, such as, if lines are drawn from its ends to any point in the circumference, they will meet at a right angle. The Q of a Scheme is as definite a phrase as the Diameter of a Circle, but we cannot replace Q in that phrase by the words Prob. Deviation, and speak of the Prob. Deviation of a Scheme, without doing some violence to language. We should have to express ourselves from another point of view, and at much greater length, and say " the Prob. Deviation of any, as yet unknown measure in the Scheme, from the Mean of all the measures from which the Scheme was constructed."

The primary idea of Q has no reference to the existence of a Mean value from which Deviations take place. It is half the difference between the measures found at the 25th and 75th Centesimal Grades. In this definition there is not the slightest allusion, direct or indirect, to the measure at the 50th Grade, which is the value of M. It is perfectly true that the measure at Grade 25° is M—Q, and that at Grade 75° is M + Q, but all this is superimposed upon the primary conception. Q stands essentially on its own basis, and has nothing to do with M. It will often happen that we shall have to deal with Prob: Deviations, but that is no reason why we should not use Q whenever it suits our purposes better, especially as statistical statements tend to be so cumbrous that every abbreviation is welcome.

The stage to which we have now arrived is this. It has been shown that the distribution of very different human qualities and faculties is approximately Normal, and it is inferred that with reasonable precautions we may treat them as if they were wholly so, in order to obtain approximate results. We shall thus deal with an entire Scheme of Deviations in terms of its Q, and with an entire Scheme of Measures in terms of its M and Q, just as we deal with an entire Circle in terms of its

radius, or with an entire Ellipse in terms of its major
and minor axes.   We can also apply the various beautiful properties of the Law of Frequency of Error to
the observed values of Q.   In doing so, we act like
woodsmen who roughly calculate the cubic contents of
the trunk of a tree, by measuring its length, and its girth
at either end, and submitting their measures to formulæ
that have been deduced from the properties of ideally
perfect straight lines and circles.   Their results prove
serviceable, although the trunk is only rudely straight
and circular.   I trust that my results will be yet closer
approximations to the truth than those usually arrived
at by the woodsmen.

*The value of a single Deviation at a known Grade
determines a Normal Scheme of Deviations.*—When
Normal Curves of Distribution are drawn within the
same limits, they differ from each other only in their
general slope ; and the slope is determined if the value
of the Deviation is given at any one specified Grade.
It must be borne in mind that the width of the limits
between which the Scheme is drawn, has no influence on
the values of the Deviations at the various Grades,
because the latter are proportionate parts of the base.
As the limits vary in width, so do the intervals between
the Grades.   When measuring the Deviation at a specified Grade for the purpose of determining the whole
Curve, it is of course convenient to adhere to the same
Grade in all cases.   It will be recollected that when
dealing with the observed curves a few pages back, I

used not one Grade but two Grades for the purpose,
namely 25° and 75°; but in the Normal Curve, the
plus and *minus* Deviations are equal in amount at all
pairs of symmetrical distances on either side of grade
50°; therefore the Deviation at either of the Grades 25°
or 75° is equal to Q, and suffices to define the entire
Curve.

The reason why a certain value Q' was stated a few
pages back to be equal to 1·015 Q, is that the Normal
Deviations at 20° and at 30°, (whose average we called
Q') are found in Table 8, to be 1·25 and 0·78; and
similarly those at 70° and 60°. The average of 1·25
and 0·78 is 1·015, whereas the Deviation at 25° or at
75° is 1·000.

*Two Measures at known Grades determine a Normal
Scheme of Measures.*—If we know the value of M as
well as that of Q we know the entire Scheme. M expresses the mean value of all the objects contained in
the group, and Q defines their variability. But if we
know the Measures at any two specified Grades, we can
deduce M and Q from them, and so determine the entire
Scheme. The method of doing this is explained in the
foot-note.[1]

[1] The following is a fuller description of the propositions in this and
in the preceding paragraph :—

(1) In any Normal Scheme, and therefore approximately in an observed
one, if the value of the Deviation is given at any *one* specified Grade the
whole Curve is determined. Let D be the given Deviation, and d the
tabular Deviation at the same Grade, as found in Table 8 ; then multiply
every entry in Table 8 by $\frac{D}{d}$. As the tabular value of Q is 1, it will become
changed into $\frac{D}{d}$.

The Charms of Statistics.—It is difficult to understand why statisticians commonly limit their inquiries to Averages, and do not revel in more comprehensive views. Their souls seem as dull to the charm of variety as that of the native of one of our flat English counties, whose retrospect of Switzerland was that, if its mountains could be thrown into its lakes, two nuisances would be got rid of at once. An Average is but a solitary fact, whereas if a single other fact be added to it, an entire Normal Scheme, which nearly corresponds to the observed one, starts potentially into existence.

Some people hate the very name of statistics, but I find them full of beauty and interest. Whenever they are not brutalised, but delicately handled by the higher methods, and are warily interpreted, their power of dealing with complicated phenomena is extraordinary. They are the only tools by which an opening can be cut

(2) If the Measures at any *two* specified Grades are given, the whole Scheme of Measures is thereby determined. Let A, B be the two given Measures of which A is the larger, and let a, b be the values of the tabular Deviations for the same Grades, as found in Table 8, not omitting their signs of *plus* or *minus* as the case may be.

Then the Q of the Scheme $= \pm \dfrac{A-B}{a-b}$. (*The sign of Q is not to be regarded ; it is merely a magnitude.*)

$$M = A - a\,Q \; ; \; \text{or } M = B - b\,Q.$$

Example : A, situated at Grade 55°, $= 14\cdot38$
                B, situated at Grade 5°, $= 9\cdot12$

The corresponding tabular Deviations are :—$a = +0\cdot19$; $b = -2\cdot44$.

Therefore $Q = \dfrac{14\cdot38 - 9\cdot12}{0\cdot19 + 2\cdot44} = \dfrac{5\cdot26}{2\cdot63} = 2\cdot0$

$M = 14\cdot38 - 0\cdot19 \times 2 = 14\cdot0$

or $= 9\cdot12 + 2\cdot44 \times 2 = 14\cdot0$

through the formidable thicket of difficulties that bars
the path of those who pursue the Science of man.

*Mechanical Illustration of the Cause of the Curve of
Frequency.*—The Curve of Frequency, and that of Distribution, are convertible : therefore if the genesis of either
of them can be made clear, that of the other becomes
also intelligible. I shall now illustrate the origin of the
Curve of Frequency, by means of an apparatus shown in
Fig. 7, that mimics in a very pretty way the conditions

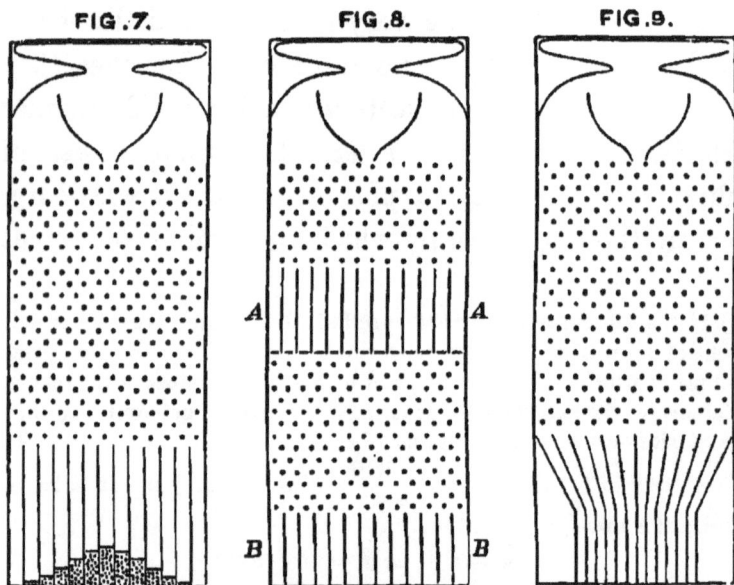

on which Deviation depends. It is a frame glazed in
front, leaving a depth of about a quarter of an inch behind the glass. Strips are placed in the upper part to act
as a funnel. Below the outlet of the funnel stand a

succession of rows of pins stuck squarely into the back-
board, and below these again are a series of vertical
compartments.   A charge of small shot is inclosed.
When the frame is held topsy-turvy, all the shot runs
to the upper end; then, when it is turned back into
its working position, the desired action commences.
Lateral strips, shown in the diagram, have the effect of
directing all the shot that had collected at the upper
end of the frame to run into the wide mouth of the
funnel.   The shot passes through the funnel and issuing
from its narrow end, scampers deviously down through
the pins in a curious and interesting way; each of them
darting a step to the right or left, as the case may be,
every time it strikes a pin.   The pins are disposed in a
quincunx fashion, so that every descending shot strikes
against a pin in each successive row.   The cascade
issuing from the funnel broadens as it descends, and, at
length, every shot finds itself caught in a compartment
immediately after freeing itself from the last row of
pins.   The outline of the columns of shot that accumulate
in the successive compartments approximates to the
Curve of Frequency (Fig. 3, p. 38), and is closely of
the same shape however often the experiment is repeated.   The outline of the columns would become more
nearly identical with the Normal Curve of Frequency,
if the rows of pins were much more numerous, the shot
smaller, and the compartments narrower; also if a larger
quantity of shot was used.

The principle on which the action of the apparatus
depends is, that a number of small and independent

accidents befall each shot in its career. In rare cases, a long run of luck continues to favour the course of a particular shot towards either outside place, but in the large majority of instances the number of accidents that cause Deviation to the right, balance in a greater or less degree those that cause Deviation to the left. Therefore most of the shot finds its way into the compartments that are situated near to a perpendicular line drawn from the outlet of the funnel, and the Frequency with which shots stray to different distances to the right or left of that line diminishes in a much faster ratio than those distances increase. This illustrates and explains the reason why mediocrity is so common.

If a larger quantity of shot is put inside the apparatus, the resulting curve will be more humped, but one half of the shot will still fall within the same distance as before, reckoning to the right and left of the perpendicular line that passes through the mouth of the funnel. This distance, which does not vary with the quantity of the shot, is the " Prob : Error," or " Prob : Deviation," of any single shot, and has the same value as our Q. But a Scheme of Frequency is unsuitable for finding the values of either M or Q. To do so, we must divide its strangely shaped *area* into four equal parts by vertical lines, which is hardly to be effected except by a tedious process of "Trial and Error." On the other hand M and Q can be derived from Schemes of Distribution with no more trouble than is needed to divide a *line* into four equal parts.

F

Order in Apparent Chaos.—I know of scarcely any-. thing so apt to impress the imagination as the wonderful form of cosmic order expressed by the " Law of Fre- quency of Error." The law would have been personified by the Greeks and deified, if they had known of it. It reigns with serenity and ·in complete self-effacement amidst the wildest confusion. The huger the mob, and the greater the apparent anarchy, the more perfect is its sway. It is the supreme law of Unreason. Whenever a large sample of chaotic elements are taken in hand and marshalled in the order of their magnitude, an un- suspected and most beautiful form of regularity proves to have been latent all along. The tops of the mar- shalled row form a flowing curve of invariable pro- portions ; and each element, as it is sorted into place, finds, as it were, a pre-ordained niche, accurately adapted to fit it. If the measurement at any two specified Grades in the row are known, those that will be found at every other Grade, except towards the extreme. ends, can be predicted in the way already explained, and with much precision.

Problems in the Law of Error.—All the properties of the Law of Frequency of Error can be expressed in terms of Q, or of the Prob: Error, just as those of a circle can be expressed in terms of its radius. The visible Schemes are not, however, to be removed too soon from our imagination. It is always well to retain a clear geometric view of the facts when we are dealing with statistical problems, which abound with dangerous pitfalls, easily overlooked by the unwary, while they are cantering gaily along upon their arithmetic. The Laws of Error are beautiful in themselves and exceedingly fascinating to inquirers, owing to the thoroughness and simplicity with which they deal with masses of materials that appear at first sight to be entanglements on the largest scale, and of a hopelessly confused description. I will mention five of the laws.

(1) The following is a mechanical illustration of the first of them. In the apparatus already described, let q stand for the Prob: Error of any one of the shots that are dispersed among the compartments BB at its base. Now cut the apparatus in two parts, horizontally through the rows of pins. Separate the parts and interpose a row of vertical compartments AA, as in Fig. 8, p. 63, where the bottom compartments, BB, corresponding to those shown in Fig. 7, are reduced to half their depth, in order to bring the whole figure within the same sized outline as before. The compartments BB are still deep enough for their purpose. It is clear that the inter-polation of the AA compartments can have no ultimate effect on the final dispersion of the shot into those at BB. Now close the bottoms of all the AA compart-ments; then the shot that falls from the funnel will be retained in them, and will be comparatively little dis-persed. Let the Prob: Error of a shot in the AA com-partments be called a. Next, open the bottom of any one of the AA compartments; then the shot it contains will cascade downwards and disperse themselves among the BB compartments on either side of the perpendicu-

lar line drawn from its starting point, and each shot
will have a Prob: Error that we will call b. Do this
for all the AA compartments in turn ; b will be the
same for all of them, and the final result must be to re-
produce the identically same system in the BB compartments that was shown in Fig. 7, and in which each
shot had a Prob: Error of q.

The dispersion of the shot at BB may therefore be
looked upon as compounded of two superimposed and
independent systems of dispersion. In the one, when
acting singly, each shot has a Prob: Error of a; in
the other, when acting singly, each shot has a Prob:
Error of b, and the result of the two acting together is
that each shot has a Prob: Error of q. What is the
relation between a, b, and q? Calculation shows that
$q^2 = a^2 + b^2$. In other words, q corresponds to the hypo-
thenuse of a right-angled triangle of which the other two
sides are a and b respectively.

(2) It is a corollary of the foregoing that a system Z,
in which each element is the Sum of a couple of independent Errors, of which one has been taken at random
from a Normal system A and the other from a Normal
system B, will itself be Normal.[1] Calling the Q of the Z
system q, and the Q of the A and B systems respectively,
a and b, then $q^2 = a^2 + b^2$.

---

[1] We may see the rationale of this corollary if we invert part of the
statement of the problem. Instead of saying that an A element deviates
from its M, and that a B element also deviates independently from its M, we
may phrase it thus : An A element deviates from its M, and its M deviates
from the B element. Therefore the deviation of the B element from the
A element is compounded of two independent deviations, as in Problem 1.

(3) Suppose that a row of compartments, whose upper
openings are situated like those in Fig. 7, page 63, are
made first to converge towards some given point below,
but that before reaching it their sloping course is
checked and they are thenceforward allowed to drop
vertically as in Fig. 9. The effect of this will be to
compress the heap of shot laterally; its outline will still
be a Curve of Frequency, but its Prob: Error will be
diminished.

The foregoing three properties of the Law of Error are
well known to mathematicians and require no demonstration here, but two other properties that are not
familiar will be of use also; proofs of them by Mr. J.
Hamilton Dickson are given in Appendix B. They are
as follows. I purposely select a different illustration to
that used in the Appendix, for the sake of presenting
the same general problem under more than one of its
applications.

(4) Bullets are fired by a man who aims at the centre
of a target, which we will call its M, and we will suppose
the marks that the bullets make to be painted red, for
the sake of distinction. The system of lateral deviations
of these red marks from the centre M will be approximately Normal, whose Q we will call c. Then another
man takes aim, not at the centre of the target, but at
one or other of the red marks, selecting these at random.
We will suppose his shots to be painted green. The
lateral distance of any green shot from the red mark
at which it was aimed will have a Prob: Error that we

will call b.   Now, if the lateral distance of a particular green mark from M is given, what is the *most probable* distance from M of the red mark at which it was aimed ?

It is $\sqrt{\dfrac{c^2}{c^2 + b^2}}$.

(5) What is the Prob: Error of this determination ? In other words, if estimates have been made for a great many distances founded upon the formula in (4), they would be correct on the average, though erroneous in particular cases.   The errors thus made would form a normal system whose Q it is desired to determine.   Its value is $\dfrac{bc}{\sqrt{(b^2 + c^2)}}$.

By the help of these five problems the statistics of heredity become perfectly manageable.   It will be found that they enable us to deal with Fraternities, Populations, or other Groups, just as if they were units. The largeness of the number of individuals in any of our groups is so far from scaring us, that they are actually welcomed as making the calculations more sure and none the less simple.

CHAPTER VI.

DATA.

Records of Family Faculties, or R. F. F. data.—Special Data.—Measures at my Anthropometric Laboratory.—Experiments on Sweet Peas.

I HAD to collect all my data for myself, as nothing existed, so far as I know, that would satisfy even my primary requirement. This was to obtain records of at least two successive generations of some population of considerable size. They must have lived under conditions that were of a usual kind, and in which no great varieties of nurture were to be found. Natural selection must have had little influence on the characteristics that were to be examined. These must be measurable, variable, and fairly constant in the same individual. The result of numerous inquiries, made of the most competent persons, was that I began my experiments many years ago on the seeds of sweet peas, and that at the present time I am breeding moths, as will be explained in a later chapter, but this book refers to a human population, which, take it all in all, is the easiest to work with when the data are once obtained,

to say nothing of its being more interesting by far than
one of sweet peas or of moths.

Record of Family Faculties, or R.F.F. Data.—The
source from which the larger part of my data is derived
consists of a valuable collection of "Records of Family
Faculties," obtained through the offer of prizes. They
have been much tested and cross-tested, and have borne
the ordeal very fairly, so far as it has been applied. It
is well to reprint the terms of the published offer, in
order to give a just idea of the conditions under which
they were compiled. It was as follows:

"Mr. Francis Galton offers 500*l*. in prizes to those
British Subjects resident in the United Kingdom who
shall furnish him before May 15, 1884, with the best
Extracts from their own Family Records.

"These Extracts will be treated as confidential documents, to be used for statistical purposes only, the
insertion of names of persons and places being required
solely as a guarantee of authenticity and to enable Mr.
Galton to communicate with the writers in cases where
further question may be necessary.

"The value of the Extracts will be estimated by the
degree in which they seem likely to facilitate the scientific investigations described in the preface to the
'Record of Family Faculties.'

"More especially:

"(*a*) By including every direct ancestor who stands
within the limits of kinship there specified.

"(*b*) By including brief notices of the brothers and

sisters (if any) of each of those ancestors. (Importance will be attached both to the completeness with which each family of brothers and sisters is described, and also to the number of persons so described.)

" (c) By the character of the evidence upon which the information is based.

" (d) By the clearness and conciseness with which the statements and remarks are made.

" The Extracts must be legibly entered either in the tabular forms contained in the copy of the 'Record of Family Faculties' (into which, if more space is wanted, additional pages may be stitched), or they may be written in any other book with pages of the same size as those of the Record, provided that the information be arranged in the same tabular form and order. (It will be obvious that uniformity in the arrangement of documents is of primary importance to those who examine and collate a large number of them.)

" Each competitor must furnish the name and address of a referee of good social standing (magistrate, clergyman, lawyer, medical practitioner, &c.), who is personally acquainted with his family, and of whom inquiry may be made, if desired, as to the general trustworthiness of the competitor.

" The Extracts must be sent prepaid and by post, addressed to Francis Galton, 42 Rutland Gate, London, S.W. It will be convenient if the letters 'R.F.F.' (Record of Family Faculties) be written in the left-hand corner of the parcel, below the address.

" The examination will be conducted by the donor of the prizes, aided by competent examiners.

" The value of the individual prizes cannot be fixed beforehand. No prize will, however, exceed 50*l.*, nor be less than 5*l.*, and 500*l.* will on the whole be awarded.

" A list of the gainers of the prizes will be posted to each of them. It will be published in one or more of the daily newspapers, also in at least one clerical, and one medical Journal."

The number of Family Records sent in reply to this offer, that deserved to be seriously considered before adjudging the prizes, barely reached 150 ; 70 of these being contributed by males, 80 by females. The remainder were imperfect, or they were marked " not for competition," but at least 10 of these have been to some degree utilised. The 150 Records were contributed by persons of very various ranks. After classing the female writers according to the profession of their husbands, if they were married, or according to that of their fathers, if they were unmarried, I found that each of the following 7 classes had 20 or somewhat fewer representatives : (1) Titled persons and landed gentry ; (2) Army and Navy; (3) Church (various denominations) ; (4) Law; (5) Medicine; (6) Commerce, higher class ; (7) Commerce, lower class. This accounts for nearly 130 of the writers of the Records ; the remainder are land agents, farmers, artisans, literary men, schoolmasters, clerks, students, and one domestic servant in a family of position.

Three cases occurred in which the Records sent by different contributors overlapped. The details are complicated, and need not be described here, but the result is that five persons have been adjudged smaller prizes than they individually deserved.

Every one of the replies refers to a very large number of persons, as will easily be understood if the fact is borne in mind that each individual has 2 parents, 4 grandparents, and 8 great parents; also that he and each of those 14 progenitors had usually brothers and sisters, who were included in the inquiry. The replies were unequal in merit, as might have been expected, but many were of so high an order that I could not justly select a few as recipients of large prizes to the exclusion of the rest. Therefore I divided the sum into two considerable groups of small prizes, all of which were well deserved, regretting much that I had none left to award to a few others of nearly equal merit to some of those who had been successful. The list of winners is reproduced below, the four years that have elapsed have of course made not a few changes in the addresses, which are not noticed here.

LIST OF AWARDS.

A PRIZE OF £7 WAS AWARDED TO EACH OF THE 40 FOLLOWING CONTRIBUTORS.

Amphlett, John, Clent, Stourbridge ; Batchelor, Mrs. Jacobstow Rectory, Stratton, N. Devon ; Bathurst, Miss K., Vicarage, Biggleswade, Bedfordshire ; Beane, Mrs. C. F., 3 Portland Place, Venner Road, Sydenham ; Berisford, Samuel, Park Villas, Park Lane, Macclesfield ; Carruthers, Mrs., Brightside, North Finchley ; Carter, Miss Jessie E., Hazelwood, The Park, Cheltenham ; Cay, Mrs., Eden House, Holyhead ; Clark, J. Edmund,

Feversham Terrace, York; Cust, Lady Elizabeth, 13 Eccleston Square, S.W.; Fry, Edward, Portsmouth, 5 The Grove, Highgate, N.; Gibson, G. A., M.D., 1 Randolph Cliff, Edinburgh; Gidley, B. Courtenay, 17 Ribblesdale Road, Hornsey, N.; Gillespie, Franklin, M.D., 1 The Grove, Aldershot; Griffith-Boscawen, Mrs., Trevalyn Hall, Wrexham; Hardcastle, Henry, 38 Eaton Square, S.W.; Harrison, Miss Edith, 68 Gloucester Place, Portman Square, W.; Hobhouse, Mrs. 4 Kensington Square, W.; Holland, Miss, Ivymeath, Snodland, Kent; Hollis, George, Dartmouth House, Dartmouth Park Hill, N.; Ingram, Mrs. Ades, Chailey, Lewis, Sussex; Johnstone, Miss C. L., 3 Clarendon Place, Leamington; Lane-Poole, Stanley, 6 Park Villas East, Richmond, Middlesex; Leathley, D. W. B., 59 Lincoln's Inn Fields (in trust for a competitor who desires her name not to be published); Lewin, Lieutenant-Colonel T. H., Colway Lodge, Lyme Regis; Lipscomb, R. H., East Budleigh, Budleigh Salterton, Devon; Malden, Henry C., Windlesham House, Brighton; Malden, Henry Elliot, Kitland, Holmwood, Surrey; McCall, Hardy Bertram, 5 St. Augustine's Road, Edgbaston, Birmingham; Moore, Miss Georgina M., 45 Chepstow Place, Bayswater, W.; Newlands, Mrs., Raeden, near Aberdeen; Pearson, David R., M.D., 23 Upper Phillimore Place, Kensington, W.; Pearson, Mrs., The Garth, Woodside Park, North Finchley: Pechell, Hervey Charles, 6 West Chapel Street, Curzon Street, W.; Roberts, Samuel, 21 Roland Gardens, S.W.; Smith, Mrs. Archibald, Riverbank, Putney, S.W.; Strachey, Mrs. Fowey Lodge, Clapham Common, S.W.; Sturge, Miss Mary C., Chilliswood, Tyndall's Park, Bristol; Sturge, Mrs. R. F., 101 Pembroke Road, Clifton; Wilson, Edward T., M.D., Westall, Cheltenham.

A PRIZE OF £5 WAS AWARDED TO EACH OF THE 44 FOLLOWING CONTRIBUTORS.

Allan, Francis J., M.D., 1 Dock Street, E.; Atkinson, Mrs., Clare College Lodge, Cambridge; Bevan, Mrs., Plumpton House, Bury St. Edmunds; Browne, Miss, Maidenwell House, Louth, Lincolnshire; Cash, Frederick Goodall, Gloucester; Chisholm, Mrs., Church Lane House, Haslemere, Surrey; Collier, Mrs. R., 7 Thames Embankment, Chelsea; Croft, Sir Herbert G. D., Lugwardine Court, Hereford; Davis, Mrs. (care of Israel Davis, 6 King's Bench Walk, Temple, E.C.); Drake, Henry H., The Firs, Lee, Kent; Ercke, J. J. G., 13, Brownhill Road, Catford, S.E.; Flint, Fenner Ludd, 83 Brecknock Road, N.; Ford, William, 4 South Square, Gray's Inn, W.C.; Foster, Rev. A. J., The Vicarage, Wootton, Bedford; Glanville-Richards, W. V. S., 23 Endsleigh Place, Plymouth; Hale, C. D. Bowditch, 8 Sussex Gardens, Hyde Park, W.; Horder, Mrs. Mark, Rothenwood, Ellen Grove, Salisbury; Jackson, Edwin, 79 Withington Road, Whalley Range, Manchester; Jackson, George, 1 St. George's Terrace, Plymouth; Kesteven, W. H., 401 Holloway Road, N.; Lawrence, Mrs.

Alfred, 16 Suffolk Square, Cheltenham ; Lawrie, Mrs., 1 Chesham Place,
S.W. ; Leveson-Gower, G. W. G., Titsey Place, Limpsfield, Surrey ; Lobb,
H. W., 66 Russell Square, W.; McConnell, Miss M. A. Brooklands,
Prestwich, Manchester ; Marshall, Mrs., Fenton Hall, Stoke-upon-Trent ;
Meyer, Mrs., 1 Rodney Place, Clifton, Bristol; Milman, Mrs., The Governor's
House, H.M. Prison, Camden Road ; Olding, Mrs. W. 4 Brunswick Road,
Brighton, Sussex ; Passingham, Mrs., Milton, Cambridge ; Pringle, Mrs.
Fairnalie, Fox Grove Road, Beckenham, Kent ; Reeve, Miss, Foxholes,
Christchurch, Hants ; Scarlett, Mrs., Boscomb Manor, Bournemouth ;
Shand, William, 57 Caledonian Road, N.; Shaw, Cecil E., Wellington
Park, Belfast ; Sizer, Miss Kate T., Moorlands, Great Huntley, Colchester ;
Smith, Miss A. M. Carter, Thistleworth, Stevenage ; Smith, Rev. Edward S.,
Viney Hall Vicarage, Blakeney, Gloucestershire ; Smith, Mrs. F. P., Cliffe
House, Sheffield ; Staveley, Edw. S. R., Mill Hill School, N.W. ; Sturge,
Miss Mary W., 17 Frederick Road, Edgbaston, Birmingham ; Terry, Mrs.,
Tostock, Bury St. Edmunds, Suffolk; Utley, W. H. Alliance Hotel,
Cathedral Gates, Manchester ; Weston, Mrs. Ensleigh, Lansdown, Bath ;
Wodehouse, Mrs. E. R. 56 Chester Square, S.W.

The material in these Records is sufficiently varied to
be of service in many inquiries. The chief subjects to
which allusion will be made in this book concern
Stature, Eye-Colour, Temper, the Artistic Faculty, and
some forms of Disease, but others are utilized that refer
to Marriage Selection and Fertility.

The following remarks in this Chapter refer almost
wholly to the data of Stature.

The data derived from the Records of Family Faculties
will be hereafter distinguished by the letters R.F.F. I
was able to extract from them the statures of 205 couples
of parents, with those of an aggregate of 930 of their
adult children of both sexes. I must repeat that when
dealing with the female statures, I transmuted them to
their male equivalents; and treated them when thus
transmuted, on equal terms with the measures of males,

except where otherwise expressed. The factor I used was 1·08, which is equivalent to adding a little less than one-twelfth to each female height. It differs slightly from the factors employed by other anthropologists, who, moreover, differ a trifle between themselves; anyhow, it suits my data better than 1·07 or 1·09. I can say confidently that the final result is not of a kind to be sensibly affected by these minute details, because it happened that owing to a mistaken direction, the computer to whom I first entrusted the figures used a somewhat different factor, yet the final results came out closely the same. These R.F.F. data have by no means the precision of the observations to be spoken of in the next paragraph. In many cases there remains considerable doubt whether the measurement refers to the height with the shoes on or off; not a few of the entries are, I fear, only estimates, and the heights are commonly given only to the nearest inch. Still, speaking from a knowledge of many of the contributors, I am satisfied that a fair share of these returns are undoubtedly careful and thoroughly trustworthy, and as there is no sign or suspicion of bias, I have reason to place confidence in the values of the Means that are derived from them. They bear the internal tests that have been applied better than might have been expected, and when checked by the data described in the next paragraph, and cautiously treated, they are very valuable.

Special Data.—A second set of data, distinguished by the name of "Special observations," concern the variations in stature among Brothers. I circulated cards of inquiry among trusted correspondents, stating that I wanted records of the heights of brothers who were more than 24 and less than 60 years of age; that it was not necessary to send the statures of all of the brothers of the same family, but only of as many of them as could be easily and accurately measured, and that the height of even two brothers would be acceptable. The blank forms sent to be filled, were ruled vertically in three parallel columns : (a) family name of each set of brothers ; (b) order of birth in each set ; (c) height without shoès, in feet and inches. A place was reserved at the bottom for the name and address of the sender. The circle of inquirers widened, but I was satisfied when I had obtained returns of 295 families, containing in the aggregate 783 brothers, some few of whom also appear in the R.F.F. data. Though these two sets of returns overlap to a trifling extent, they are practically independent. I look upon the " Special Observations " as being quite as trustworthy as could be expected in any such returns. They bear every internal test that I can apply to them in a very satisfactory manner. The measures are commonly recorded to quarter or half inches.

Measures at my Anthropometric Laboratory.—A third set of data have been incidentally of service. They are the large lists of measures, nearly 10,000 in number, made at my Anthropometric Laboratory in the International Health Exhibition of 1884.

4. *Experiments on Sweet Peas.*—The last of the data

that I need specify were the very first that I used ; they refer to the sizes of seeds, which are equivalent to the Statures of seeds.   I both measured and weighed them, but after assuring myself of the equivalence of the two methods (see Appendix C.), confined myself to ascertaining the weights, as they were much more easily ascertained than the measures.   It is more than 10 years since I procured these data.   They were the result of an extensive series of experiments on the produce of seeds of different sizes, but of the same species, conducted for the following reasons. I had endeavoured to find a population possessed of some measurable characteristic that was suitable for investigating the causes of the statistical similarity between successive generations of a people, as will hereafter be discussed in Chapter VIII.   At last I determined to experiment on seeds, and after much inquiry of very competent advisers, selected sweet-peas for the purpose. They do not cross-fertilize, which is a very exceptional condition among plants ; they are hardy, prolific, of a convenient size to handle, and nearly spherical ; their weight does not alter perceptibly when the air changes from damp to dry, and the little pea at the end of the pod, so characteristic of ordinary peas, is absent in sweet-peas.   I began by weighing thousands of them individually, and treating them as a census officer would treat a large population.   Then I selected with great pains several sets for planting.   Each set contained seven little packets, numbered K, L, M, N, O, P, and Q, each of the seven packets contained ten seeds of almost exactly the same weight; those in K being the heaviest, L the next heaviest, and so down to Q, which was the lightest. The precise weights are given in Appendix C, together with the corresponding diameters, which I ascertained by laying 100 peas of the same weight in a row. The weights run in an arithmetic series, having a common average difference of 0·172 grain. I do not of course profess to work to thousandths of a grain, though I did work to somewhat less than one hundredth of a grain; therefore the third decimal place represents little more than an arithmetical working value which has to be regarded in multiplications, lest an error of sensible importance should be introduced by its neglect. Curiously enough, the diameters were found also to run approximately in an arithmetic series, owing, I suppose, to the misshape and corrugations of the smaller seeds, which gave them a larger diameter than if they had been plumped out into spheres. All this is shown in the Appendix, where it will be seen that I was justified in sorting the seeds by the convenient method of the balance and weights, and of accepting the weights as directly proportional to the mean diameters.

In each experiment, seven beds were prepared in parallel rows; each was 1½ feet wide and 5 feet long. Ten holes of 1 inch deep were dibbled at equal distances apart along each bed, and a single seed was put into each hole. The beds were then bushed over to keep off the birds. Minute instructions were given to ensure uniformity, which I need not repeat here. The end of all was that the seeds as they became ripe were

collected from time to time and put into bags that I
had sent, lettered from K to Q, the same letters having
been stuck at the ends of the beds. When the crop was
coming to an end, the whole remaining produce of each
bed, including the foliage, was torn up, tied together,
labelled, and sent to me. Many friends and acquaint-
ances had each undertaken the planting and culture of
a complete set, so that I had simultaneous experiments
going on in various parts of the United Kingdom from
Nairn in the North to Cornwall in the South. Two
proved failures, but the final result was that I obtained
the more or less complete produce of seven sets; that is
to say, the produce of $7 \times 7 \times 10$, or of 490 carefully
weighed parent seeds. Some additional account of the
results is given in Appendix C.

It would be wholly out of place to enter here into
further details of the experiments, or to narrate the
numerous little difficulties and imperfections I had to
contend with, and how I balanced doubtful cases ; how
I divided returns into groups to see if they confirmed
one another, or how I conducted any other well-known
statistical operation. Suffice it to say that I took immense pains, which, if I had understood the general
conditions of the problem as clearly as I do now, I
should not perhaps have cared to bestow. The results
were most satisfactory. They gave me two data, which
were all that I wanted in order to understand in its
simplest approximate form, the way in which one
generation of a people is descended from a previous one ;
and thus I got at the heart of the problem at once.

CHAPTER VII.

DISCUSSION OF THE DATA OF STATURE.

Stature as a subject for inquiry.—Marriage Selection.—Issue of unlike Parents.—Description of the Tables of Stature. Mid-Stature of the Population.—Variability of the Population.—Variability of Mid-Parents.—Variability in Co-Fraternities.—Regression : *a*, Filial ; *b*, Mid-Parental ; *c*, Parental ; *d*, Fraternal.—Squadrons of Statures.—Successive Generations of a People.—Natural Selection.—Variability in Fraternities.—Trustworthiness of the Constants.—General view of Kinship.—Separate Contribution from each Ancestor.—Pedigree Moths.

Stature as a Subject for Inquiry.—The first of these inquiries into the laws of human heredity deals with hereditary Stature, which is an excellent subject for statistics. Some of its merits are obvious enough, such as the case and frequency with which it may be measured, its practical constancy during thirty-five or forty years of middle life, its comparatively small dependence upon differences of bringing up, and its inconsiderable influence on the rate of mortality. Other advantages which are not equally obvious are equally great. One of these is due to the fact that human stature is not a simple element, but a sum of the accumulated lengths or

G 2

thicknesses of more than a hundred bodily parts, each
so distinct from the rest as to have earned a name by
which it can be specified. The list includes about fifty
separate bones, situated in the skull, the spine, the
pelvis, the two legs, and in the two ankles and feet.
The bones in both the lower limbs have to be counted,
because the Stature depends upon their average length.
The two cartilages interposed between adjacent bones,
wherever there is a movable joint, and the single
cartilage in other cases, are rather more numerous than
the bones themselves. The fleshy parts of the scalp
of the head and of the soles of the feet conclude the
list Account should also be taken of the shape and
set of the many bones which conduce to a more or less
arched instep, straight back, or high head. I noticed
in the skeleton of O'Brien, the Irish giant, at the College
of Surgeons, which is the tallest skeleton in any English
museum, that his great stature of about 7 feet 7 inches
would have been a trifle increased if the faces of his
dorsal vertebræ had been more parallel than they are,
and his back consequently straighter. ·

This multiplicity of elements, whose variations are to
some degree independent of one another, some tending
to lengthen the total stature, others to shorten it,
corresponds to an equal number of sets of rows of
pins in the apparatus Fig. 7, p. 63, by which the cause
of variability was illustrated. The larger the number of
these variable elements, the more nearly does the variability of their sum assume a "Normal" character, though
the approximation increases only as the square root of

their number. The beautiful regularity in the Statures of a population, whenever they are statistically marshalled in the order of their heights, is due to the number of variable and quasi-independent elements of which Stature is the sum.

Marriage Selection.—Whatever may be the sexual preferences for similarity or for contrast, I find little indication in the average results obtained from a fairly large number of cases, of any single measurable personal peculiarity, whether it be stature, temper, eye-colour, or artistic tastes, in influencing marriage selection to a notable degree. Nor is this extraordinary, for though people may fall in love for trifles, marriage is a serious act, usually determined by the concurrence of numerous motives. Therefore we could hardly expect either shortness or tallness, darkness or lightness in complexion, or any other single quality, to have in the long run a large separate influence.

I was certainly surprised to find how imperceptible was the influence that even good and bad Temper seemed to exert on marriage selection. A list was made (see Appendix D) of the observed frequency of marriages between persons of each of the various classes of Temper, in a group of 111 couples, and I calculated what would have been the relative frequency of intermarriages between persons of the various classes, if the same number of males and females had been paired at random. The result showed that the observed list agreed closely with the calculated list, and therefore that these observations gave no evidence of discriminative selection in respect to Temper. The good-tempered husbands were 46 per cent. in number, and, between them, they married 22 good-tempered and 24 bad-tempered wives; whereas calculation, having regard to the relative proportions of good and bad Temper in the two sexes, gave the numbers as 25 and 21. Again, the bad-tempered husbands, who were 54 per cent. in number, married 31 good-tempered and 23 bad-tempered wives, whereas calculation gave the number as 30 and 24. This rough summary is a just expression of the results arrived at by a more minute analysis, which is described in the Appendix, and need not be repeated here.

Similarly as regards Eye-Colour. If we analyse the marriages between the 78 couples whose eye-colours are described in Chapter VIII., and compare the observed results with those calculated on the supposition that Eye-Colour has no influence whatever in marriage selection, the two lists will be found to be much alike. Thus where both of the parents have eyes of the same colour, whether they be light, or hazel, or dark, the percentage results are almost identical, being 37, 3, and 8 as observed, against 37, 2, and 7 calculated. Where one parent is hazel-eyed and the other dark-eyed, the marriages are as 5 observed against 7 calculated. But the results run much less well together in the other two possible combinations, for where one parent is light and the other hazel-eyed, they give 23 observed against 15 calculated; and where one parent is light and the other dark-eyed, they give 24 observed against 32 calculated.

The effect of Artistic Taste on marriage selection is discussed in Chapter X., and this also is shown to be small. The influence on the race of Bias in Marriage Selection will be discussed in that chapter.

I have taken much trouble at different times to determine whether Stature plays any sensible part in marriage selection. I am not yet prepared to offer complete results, but shall confine my remarks for the present to the particular cases with which we are now concerned. The shrewdest test is to proceed under the guidance of Problem 2, page 68. I find the Q of Stature among the male population to be 1·7 inch, and similarly for the transmuted statures of the female population. Consequently if the men and (transmuted) women married at random so far as stature was concerned, the Q in a group of couples, each couple consisting of a pair of *summed* statures, would be $\sqrt{2} \times 1\cdot7$ inches $= 2\cdot41$ inches. Therefore the Q in a group of which each element is the *mean* stature of a couple, would be half that amount, or 1·20 inch. This closely corresponds to what I derived from the data contained in the first and in the last column but one of Table 11. The word " Mid-Parent," in the headings to those columns, expresses an ideal person of composite sex, whose Stature is half way between the Stature of the father and the transmuted Stature of the mother. I therefore conclude that marriage selection does not pay such regard to Stature, as deserves being taken into account in the cases with which we are concerned.

I tried the question in another but ruder way, by

dividing (see Table 9) the male and female parents respectively into three nearly equal groups, of tall, medium,
and short.   It was impracticable to make them precisely
equal, on account of the roughness with which the
measurements were recorded, so I framed rules that
seemed best adapted to the case.   Consequently the
numbers of the tall and short proved to be only approximately and not exactly equal, and the two together
were only approximately equal to the medium cases.
The final results were :—32 instances where one parent
was short and the other tall, and 27 where both were
short or both were tall.   In other words, there were 32
cases of contrast in marriage, to 27 cases of likeness.
I do not regard this difference as of consequence,
because the numbers are small, and because a slight
change in the limiting values assigned to shortness and
tallness, would have a sensible effect upon the result.
I am therefore content to ignore it, and to regard the
Statures of married folk just as if their choice in marriage had been wholly independent of stature.   The
importance of this supposition in facilitating calculation
will be appreciated as we proceed.

Issue of Unlike Parents.—We will next discuss the
question whether the Stature of the issue of unlike
parents betrays any notable evidence of their unlikeness,
or whether the peculiarities of the children do not rather
depend on the *average* of two values; one the Stature
of the father, and the other the transmuted Stature
of the mother; in other words, on the Stature of

that ideal personage to whom we have already been introduced, under the name of a Mid-Parent. Stature has already been spoken of as a well-marked instance of the heritages that blend freely in the course of hereditary transmission. It now becomes necessary to substantiate the statement, because it is proposed to trace the relationship between the Mid-Parent and the Son. It would not be possible to discuss the relationship between either parent singly, and the son, in a trustworthy way, without the help of a much larger number of observations than are now at my disposal. They ought to be numerous enough to give good assurance that the cases of tall and short, among the unknown parents, shall neutralise one another; otherwise the uncertainty of the stature of the unknown parent would make the results uncertain to a serious degree. I am heartily glad that I shall be able fully to justify the method of dealing with Mid-Parentages instead of with single Parents.

The evidence is as follows :—If the Stature of children depends only upon the *average* Stature of their two Parents, that of the mother having been first transmuted, it will make no difference in a Fraternity whether one of the Parents was tall and the other short, or whether they were alike in Stature. But if some children resemble one Parent in Stature and others resemble the other, the Fraternity will be more diverse when their Parents had differed in Stature than when they were alike. We easily acquaint ourselves with the facts by separating a considerable number of Fraternities into two contrasted groups: (a) those who are the progeny

of Like Parents; (b) those who are the progeny of
Unlike Parents. Next we write the statures of the
individuals in each Fraternity under the form of
$M + (\pm D)$ (see page 51), where M is the mean stature
of the Fraternity, and D is the deviation of any one of
its members from M. Then we marshal all the values
of D that belong to the group a, into one Scheme of
deviations, and all those that belong to the group b
into another Scheme, and we find the Q of each. If it
should be the same, then there is no greater diversity
in the a Group than there is in the b Group, and such
proves to be the case. I applied the test (see Table 10)
to a total of 525 children, and found that they were no
more diverse in the one case than in the other. I
therefore conclude that we have only to look to the
Stature of the Mid-Parent, and need not care whether
the Parents are or are not unlike one another.

The advantages of Stature as a subject from which the
simple laws of heredity may be studied, will now be
well appreciated. It is nearly constant in the same
adult, it is frequently measured and recorded; its discussion need not be entangled with considerations of
marriage selection. It is sufficient to consider the Stature
of the Mid-Parent and not those of the two Parents
separately. Its variability is Normal, so that much use
may be made of the curious properties of the law of
Frequency of Error in cross-testing the several conclusions, and I may add that in all cases they have
borne the test successfully.

The only drawback to the use of Stature in statistical inquiries, is its small variability, one half of the population differing less than 1·7 inch from the average of all of them. In other words, its Q is only 1·7 inch.

Description of the Tables of Stature.—I have arranged and discussed my materials in a great variety of ways, to guard against rash conclusions, but do not think it necessary to trouble the reader with more than a few Tables, which afford sufficient material to determine the more important constants in the formulæ that will be used.

Table 11, R.F.F., refers to the relation between the Mid-Parent and his (or should we say *its* ?) Sons and Transmuted Daughters, and it records the Statures of 928 adult offspring of 205 Mid-Parents. It shows the distribution of Stature among the Sons of each successive group of Mid-Parents, in which the latter are all of the same Stature, reckoning to the nearest inch. I have calculated the M of each line, chiefly by drawing Schemes from the entries in it. Their values are printed at the ends of the lines and they form the right-hand column of the Table.

Tables 12 and 13 refer to the relation between Brothers. The one is derived from the R.F.F. and the other from the Special data. They both deal with small or moderately sized Fraternities, excluding the larger ones for reasons that will be explained directly, but the R.F.F. Table is the least restricted in this respect, as it only excludes families of 6 brothers and upwards. The data

were so few in number that I could not well afford to lop
off more.　These Tables were constructed by registering
the differences between each possible pair of brothers in
each family : thus if there were three brothers, A, B,
and C, in a particular family, I entered the differences
of stature between A and B, A and C, and B and C.,
four brothers gave rise to 6 entries, and five brothers to
10 entries.　The larger Fraternities were omitted, as the
very large number of different pairs in them would
have overwhelmed the influence of the smaller Fraternities.　Large Fraternities are separately dealt with in
Table 14.

We can derive some of the constants by more than
one method ; and it is gratifying to find how well the
results of different methods confirm one another.

Mid-Stature of the Population.—The Median, Mid-
Stature, or **M** of the general Population is a value of
primary importance in this inquiry.　Its value will be
always designated by the symbol P, and it may be
deduced from the bottom lines of any one of the three
Tables.　I obtain from them respectively the values
68·2, *68·5*, 68·4, but the middle of these, which is
printed in italics, is a smoothed result.　It is one of the
only two smoothed values in the whole of my work, and
was justifiably corrected, because the observed values
that happen to lie nearest to the Grade of 50° ran out of
harmony with the rest of the curve.　It is therefore
reasonable to consider its discrepancy as fortuitous,
although it amounts to more than 0·15 inch.　The

series in question refers to R.F.F. brothers, who, owing to the principle on which the Table is constructed, are only a comparatively small sample taken out of the R.F.F. Population, and on a principle that gave greater weight to a few large families than to all the rest. Therefore it could not be expected to give rise to so regular a Scheme for the general R.F.F. Population as Table 11, which was fairly based upon the whole of it. Less accuracy was undoubtedly to have been expected in this group than in either of the others.

Variability of the Population.—The value of Q in the Statures of the general Population is to be deduced from the bottom lines of any one of the Tables 11, 12, and 13. The three values of it that I so obtain, are 1·65, 1·7, and 1·7 inch. I should mention that the method of the treatment originally adopted, happened also to make the first of these values 1·7 inch, so I have no hesitation in accepting 1·7 as the value for all my data.

Variability of Mid-Parents.—The value of Q in a Scheme drawn from the Statures of the R.F.F. Mid-Parents according to the data in Table 11, is 1·19 inches. Now it has already been shown that if marriage selection is independent of stature, the value of Q in the Scheme of Mid-parental Statures would be equal to its value in that of the general Population (which we have just seen to be 1·7 inch), divided by the square root of 2 ; that is by 1·45. This calculation makes it to be

1·21 inch, which agrees excellently with the observed value.[1]

Variability in Co-Fraternities.—As all the Adult Sons and Transmuted Daughters of the *same* Mid-Parent, form what is called a Fraternity, so all the Adult Sons and Transmuted Daughters of a *group* of Mid-Parents who have the same Stature (reckoned to the nearest inch) will be termed a Co-Fraternity. Each line in Table 11 refers to a separate Co-Fraternity and expresses the distribution of Stature among them. There are three reasons why Co-Fraternals should be more diverse among themselves than brothers. First, because their Mid-Parents are not of identical height, but may differ even as much as one inch. Secondly, because their grandparents, great-grandparents, and so on indefinitely backwards, may have differed widely. Thirdly, because the nurture or rearing of Co-Fraternals is more various than that of Fraternals. The brothers in a Fraternity of townsfolk do not seem to differ more among themselves than those in a Fraternity of country-folk, but a mixture of Fraternities derived indiscriminately from the two sources, must show greater diversity than either of them taken by themselves. The large differences between town and country-folk, and those between persons of different social classes, are conspicuous in the data contained in the Report of the

---

[1] In all my values referring to human stature, the second decimal is rudely approximate. I am obliged to use it, because if I worked only to tenths of an inch, sensible errors might creep in entirely owing to arithmetical operations.

Anthropological Committee to the British Association in 1880, and published in its Journal.

I concluded after carefully studying the chart upon which each of the individual observations from which Table 11 was constructed, had been entered separately in their appropriate places, and not clubbed into groups as in the Tables, that the value of Q in each Co-Fraternal group was roughly the same, whatever their Mid-Parental value might have been. It was not quite the same, being a trifle larger when the Mid-Parents were tall than when they were short. This justifies what will be said in Appendix E about the Geometric Mean; it also justifies neglect in the present inquiry of the method founded upon it, because the improvement in the results to which it might lead, would be insignificant, while its use would have added to the difficulty of explanation, and introduced extra trouble throughout, to the reader more than to myself. The value that I adopt for Q in every Co-Fraternal group, is 1·5 inch.

Regression.—a. Filial: However paradoxical it may appear at first sight, it is theoretically a necessary fact, and one that is clearly confirmed by observation, that the Stature of the adult offspring must on the whole, be more *mediocre* than the stature of their Parents; that is to say, more near to the M of the general Population. Table 11 enables us to compare the values of the M in different Co-Fraternal groups with the Statures of their respective Mid-Parents. Fig. 10 is a graphical representation of the meaning of

the Table so far as it now concerns us. The horizontal
dotted lines and the graduations at their sides, correspond to the similarly placed lines of figures and
graduations in Table 11. The dot on each line shows
the point where its M falls. The value of its M is to
be read on the graduations along the top, and is the
same as that which is given in the last column of
Table 11. It will be perceived that the line drawn

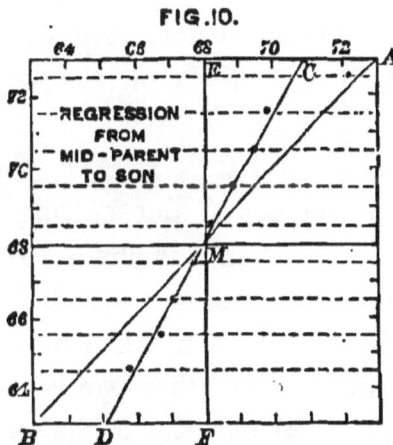

FIG. 10.

through the centres of the dots, admits of being interpreted by the straight line C D, with but a small
amount of give and take; and the fairness of this
interpretation is confirmed by a study of the MS. chart
above mentioned, in which the individual observations
were plotted in their right places.

Now if we draw a line A B through every point where
the graduations along the top of Fig. 10, are the same
as those along the sides, the line will be straight and
will run diagonally. It represents what the Mid-

Statures of the Sons would be, if they were on the average identical with those of their Mid-Parents. Most obviously A B does *not* agree with C D; therefore Sons do *not*, on the average, resemble their Mid-Parents. On examining these lines more closely, it will be observed that A B cuts C D at a point M that fairly corresponds to the value of $68\frac{1}{4}$ inches, whether its value be read on the scale at the top or on that at the side. This is the value of P, the Mid-Stature of the population. Therefore it is only when the Parents are mediocre, that their Sons on the average resemble them.

Next draw a vertical line, E M F, through M, and let E C A be any horizontal line cutting ME at E, MC at E, and MA at A. Then it is obvious that the ratio of EA to EC is constant, whatever may be the position of E C A. This is true whether E C A be drawn above or like F D B, below M. In other words, the proportion between the Mid-Filial and the Mid-Parental deviation is constant, whatever the Mid-Parental stature may be. I reckon this ratio to be as 2 to 3 : that is to say, the Filial deviation from P is on the average only two-thirds as wide as the Mid-Parental Deviation. I call this ratio of 2 to 3 the ratio of " Filial Regression." It is the proportion in which the Son is, on the average, less exceptional than his Mid-Parent.

My first estimate of the average proportion between the Mid-Filial and the Mid-Parental deviations, was made from a study of the MS. chart, and I then reckoned it as 3 to ·5. The value given above was

H

afterwards substituted, because the data seemed to admit of that interpretation also, in which case the fraction of two-thirds was preferable as being the more simple expression. I am now inclined to think the latter may be a trifle too small, but it is not worth while to make alterations until a new, larger, and more accurate series of observations can be discussed, and the whole work revised. The present doubt only ranges between nine-fifteenths in the first case and ten-fifteenths in the second.

This value of two-thirds will therefore be accepted as the amount of Regression, on the average of many cases, from the Mid-Parental to the Mid-Filial stature, whatever the Mid-Parental stature may be.

As the two Parents contribute equally, the contribution of either of them can be only one half of that of the two jointly; in other words, only one half of that of the Mid-Parent. Therefore the average Regression from the Parental to the Mid-Filial Stature must be the one half of two-thirds, or one-third. I am unable to test this conclusion in a satisfactory manner by direct observation. The data are barely numerous enough for dealing even with questions referring to Mid-Parentages; they are quite insufficient to deal with those that involve the additional large uncertainty introduced owing to an ignorance of the Stature of one of the parents. I have entered the Uni-Parental and the Filial data on a MS. chart, each in its appropriate place, but they are too scattered and irregular to make it useful to give the results in detail. They seem to show a Regression of about two-fifths, which differs from that of one-third in the ratio of 6 to 5. This direct observation is so inferior in value to the inferred result, that I disregard it, and am satisfied to adopt the value given by the latter, that is to say, of one-third, to express the average Regression from either of the Parents to the Son.

b. Mid-Parental: The converse relation to that which we have just discussed, namely the relation between the unknown stature of the Mid-Parent and the known Stature of the Son, is expressed by a fraction that is very far from being the converse of two-thirds. Though the Son deviates on the average from P only $\frac{2}{3}$ as widely as his Mid-parent, it does not in the least follow that the Mid-parent should deviate on the average from P, $\frac{3}{2}$ or $1\frac{1}{2}$, as widely as the Son. The Mid-Parent is not likely to be more exceptional than the son, but quite the contrary. The number of individuals who are nearly mediocre is so preponderant, that an exceptional man is more frequently found to be the exceptional son of mediocre parents than the average son of very exceptional parents. This is clearly shown by Table 11, where the very same observations which give the average value of Filial Regression when it is read in one way, gives that of the Mid-Parental Regression when it is read in another way, namely down the vertical columns, instead of along the horizontal lines. It then shows that the Mid-Parent of a man deviates on the average from P, only one-third as much as the man himself. This value of $\frac{1}{3}$ is four and a half times smaller than the numerical converse of $\frac{3}{2}$, since $4\frac{1}{2}$, or $\frac{9}{2}$, being multiplied into $\frac{1}{3}$, is equal to $\frac{3}{2}$.

c. Parental: As a Mid-Parental deviation is equal to one-half of the two Parental deviations, it follows that the Mid-Parental Regression must be equal to one-half of the sum of the two Parental Regressions. As the latter are equal to one another it follows that all three must have the same value. In other words, the average Mid-Parental Regression being $\frac{1}{3}$, the average Parental Regression must be $\frac{1}{3}$ also.

As there was much appearance of paradox in the above strongly contrasted results, I looked carefully into the run of the figures in Table 11. They were deduced, as already said, from a MS. chart on which the stature of every Son and the transmuted Stature of every Daughter is entered opposite to that of the Mid-Parent, the transmuted Statures being reckoned to the nearest tenth of an inch, and the position of the other entries being in every respect exactly as they were recorded. Then the number of entries in each square inch were counted, and copied in the form in which they appear in the Table. I found it hard at first to catch the full significance of the entries, though I soon discovered curious and apparently very interesting relations between them. These came out distinctly after I had "smoothed" the entries by writing at each intersection between a horizontal line and a ver-

tical one, the sum of the entries in the four adjacent
squares. I then noticed (see Fig. 11) that lines drawn
through entries of the same value formed a series of
concentric and similar ellipses. Their common centre
lay at the intersection of those vertical and horizontal
lines which correspond to the value of 68¼ inches, as
read on both the top and on the side scales. Their
axes were similarly inclined. The points where each
successive ellipse was touched by a horizontal tangent,
lay in a straight line that was inclined to the vertical in

FIG .II.

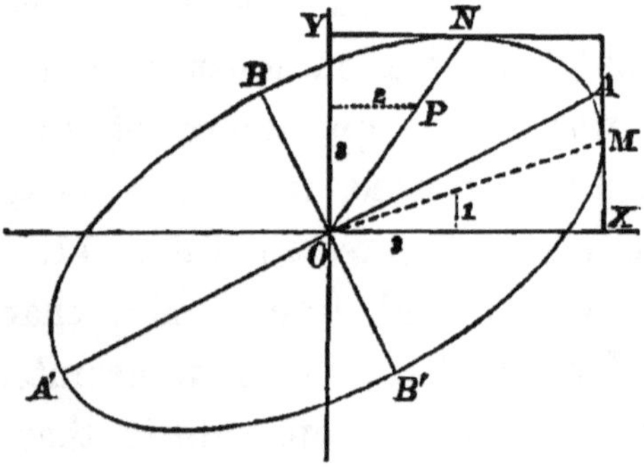

the ratio of ⅔, and those where the ellipses were touched
by a vertical tangent, lay in a straight line inclined to
the horizontal in the ratio of ⅓. It will be obvious
on studying Fig. 11 that the point where each successive horizontal line touches an ellipse is the point
at which the greatest value in the line will be found.
The same is true in respect to the successive vertical lines.
Therefore these ratios confirm the values of the Ratios
of Regression, already obtained by a different method,
namely those of ⅔ from Mid-Parent to Son, and of

$\frac{1}{3}$ from Son to Mid-Parent. These and other relations were evidently a subject for mathematical
analysis and verification. It seemed clear to me that
they all depended on three elementary measures, supposing the law of Frequency of Error to be applicable
throughout; namely (1) the value of Q in the General
Population, which was found to be 1·7 inch; (2) the
value of Q in any Co-Fraternity, which was found to be
1·5 inch; (3) the Average Regression of the Stature of
the Son from that of the Mid-Parent, which was found
to be $\frac{2}{3}$. I wrote down these values, and phrasing the
problem in abstract terms, disentangled from all reference to heredity, submitted it to Mr. J. D. Hamilton
Dickson, Tutor of St. Peter's College, Cambridge (see
Appendix B). I asked him kindly to investigate for
me the Surface of Frequency of Error that would result
from these three data, and the various shapes and other
particulars of its sections that were made by horizontal
planes, inasmuch as they ought to form the ellipses of
which I spoke.

The problem may not be difficult to an accomplished
mathematician, but I certainly never felt such a glow
of loyalty and respect towards the sovereignty and wide
sway of mathematical analysis as when his answer arrived,
confirming, by purely mathematical reasoning, my various and laborious statistical conclusions with far more
minuteness than I had dared to hope, because the data
ran somewhat roughly, and I had to smooth them with
tender caution. His calculation corrected my observed
value of Mid-Parental Regression from $\frac{1}{3}$ to $\frac{6}{17.6}$; the

relation between the major and minor axis of the ellipses was changed 3 per cent. ; and their inclination to one another was changed less than 2°.[1]

It is obvious from this close accord of calculation with observation, that the law of Error holds throughout with sufficient precision to be of real service, and that the various results of my statistics are not casual and disconnected determinations, but strictly interdependent.

I trust it will have become clear even to the most non-mathematical reader, that the law of Regression in Stature refers primarily to Deviations, that is, to measurements made from the *level of mediocrity* to the

---

[1] The following is a more detailed comparison between the calculated and the observed results. The latter are enclosed in brackets. The letters refer to Fig. 11 :—

Given—

The " Probable Error " of each system of Mid-Parentages = 1·22 inch. (This was an earlier determination of its value ; as already said, the second decimal is to be considered only as approximate.)

Ratio of mean filial regression = ⅔.

" Prob. Error" of each Co-Fraternity = 1·50 inch.

Sections of surface of frequency parallel to XY are true ellipses.

    (Obs.—Apparently true ellipses.)

$MX : YO = 6 : 17\cdot5$, or nearly $1 : 3$.

    (Obs.—1 : 3.)

Major axes to minor axes $= \sqrt{7} : \sqrt{2} = 10 : 5\cdot35$.

    (Obs.—10 : 5·1.)

Inclination of major axes to OX $= 26° 36'$.

    (Obs. 25°.)

Section of surface parallel to XZ is a true Curve of Frequency.

    (Obs.—Apparently so.)

" Prob. Error ", the Q of that curve, = 1.07 inch.

    (Obs.—1·00, or a little more.)

crown of the head, upwards or downwards as the case
may be, and not from the *ground* to the crown of the
head. (In the population with which I am now dealing,
the level of mediocrity is 68¼ inches (without shoes).)
The law of Regression in respect to Stature may be
phrased as follows; namely, that the Deviation of the
Sons from P are, on the average, equal to one-third of
the deviation of the Parent from P, and in the same
direction.   Or more briefly still:—If P + ($\pm$ D) be the
Stature of the Parent, the Stature of the offspring will
on the average be P + ($\pm \frac{1}{3}$ D).

If this remarkable law of Regression had been based
only on those experiments with seeds, in which I first
observed it, it might well be distrusted until otherwise
confirmed.   If it had been corroborated by a comparatively small number of observations on human stature,
some hesitation might be expected before its truth could
be recognised in opposition to the current belief that the
child tends to resemble its parents.   But more can be
urged than this.   It is easily to be shown that we ought
to expect Filial Regression, and that it ought to amount
to some constant fractional part of the value of the Mid-
Parental deviation.   All of this will be made clear in a
subsequent section, when we shall discuss the cause of
the curious statistical constancy in successive generations
of a large population.   In the meantime, two different
reasons may be given for the occurrence of Regression;
the one is connected with our notions of stability of
type, and of which no more need now be said; the
other is as follows:—The child inherits partly from his

parents, partly from his ancestry. In every population that intermarries freely, when the genealogy of any man is traced far backwards, his ancestry will be found to consist of such varied elements that they are indistinguishable from a sample taken at haphazard from the general Population. The Mid-Stature M of the remote ancestry of such a man will become identical with P; in other words, it will be mediocre. To put the same conclusion into another form, the most probable value of the Deviation from P, of his Mid-Ancestors in any remote generation, is zero.

For the moment let us confine our attention to some one generation in the remote ancestry on the one hand, and to the Mid-Parent on the other, and ignore all other generations. The combination of the zero Deviation of the one with the observed Deviation of the other is the combination of nothing with something. Its effect resembles that of pouring a measure of water into a vessel of wine. The wine is diluted to a constant fraction of its alcoholic strength, whatever that strength may have been.

Similarly with regard to every other generation. The Mid-Deviation in any near generation of the ancestors will have a value intermediate between that of the zero Deviation of the remote ancestry, and of the observed Deviation of the Mid-Parent. Its combination with the Mid-Parental Deviation will be as if a mixture of wine and water in some definite proportion, and not pure water, had been poured into the wine. The process throughout is one of proportionate dilutions, and the joint effect of all of them is to weaken the original alcoholic strength in a constant ratio.

The law of Regression tells heavily against the full hereditary transmission of any gift. Only a few out of many children would be likely to differ from mediocrity so widely as their Mid-Parent, and still fewer would differ as widely as the more exceptional of the two Parents. The more bountifully the Parent is gifted by nature, the more rare will be his good fortune if he begets a son who is as richly endowed as himself, and still more so if he has a son who is endowed yet more largely. But the law is even-handed; it levies an equal succession-tax on the transmission of badness as of goodness. If it discourages the extravagant hopes of a gifted parent that his children will inherit all his powers; it no less discountenances extravagant fears that they will inherit all his weakness and disease.

It must be clearly understood that there is nothing in these statements to invalidate the general doctrine that the children of a gifted pair are much more likely to be gifted than the children of a mediocre pair. They merely express the fact that the ablest of all the children of a few gifted pairs is not likely to be as gifted as the ablest of all the children of a very great many mediocre pairs.

The constancy of the ratio of Regression, whatever may be the amount of the Mid-Parental Deviation, is now seen to be a reasonable law which might have been foreseen. It is so simple in its relations that I have contrived more than one form of apparatus by which the probable stature of the children of known parents can be mechanically reckoned. Fig. 12 is a representation of one of them, that is worked with pulleys and weights. A, B, and C are three thin wheels with grooves round their edges.   They are screwed together so as to form a single piece that turns easily on its axis. The weights M and F are attached to either end of a thread that passes over the movable pulley D. The pulley itself hangs from a thread which is wrapped two or three times round the groove of B and is then secured to the wheel. The weight SD hangs from a thread that is wrapped two or three times round the groove of A, and is then secured to the wheel. The diameter of A is to that of B as 2 to 3. Lastly, a thread is wrapped in the opposite direction round the wheel C, which may have any convenient diameter; and is

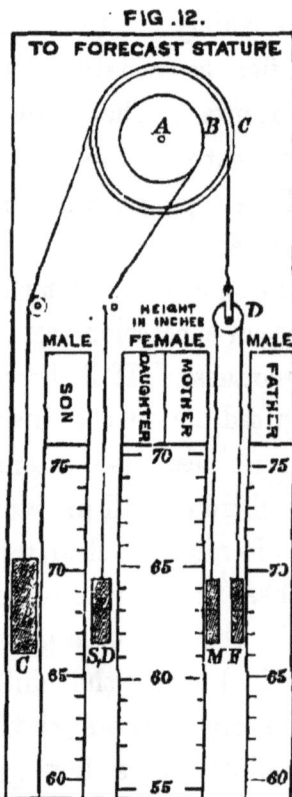

FIG .12.

TO FORECAST STATURE

attached to a counterpoise. M refers to the male statures, F to the female ones, S to the Sons, D to the Daughters.

The scale of Female Statures differs from that of the Males, each Female height being laid down in the position which would be occupied by its male equivalent.

Thus 56 is written in the position of 60·48 inches, which is equal to 56 × 1·08. Similarly, 60 is written in the position of 64·80, which is equal to 60 × 1·08.

It is obvious that raising M will cause F to fall, and *vice versâ*, without affecting the wheel AB, and therefore without affecting SD; that is to say, the Parental Differences may be varied indefinitely without affecting the Stature of the children, so long as the Mid-Parental Stature is unchanged. But if the Mid-Parental Stature is changed to any specified amount, then that of SD will be changed to ⅔ of that amount.

The weights M and F have to be set opposite to the heights of the mother and father on their respective scales; then the weight SD will show the most probable heights of a Son and of a Daughter on the corresponding scales. In every one of these cases, it is the fiducial mark in the middle of each weight by which the reading is to be made. But, in addition to this, the length of the weight SD is so arranged that it is an equal chance (an even bet) that the height of each Son or each Daughter will lie within the range defined by the upper and lower edge of the weight, on their respective scales. The length of SD is 3 inches, which is twice the Q of the Co-Fraternity; that is, 2 × 1·50 inch.

d. Fraternal : In seeking for the value of Fraternal Regression, it is better to confine ourselves to the Special data given in Table 13, as they are much more trustworthy than the R.F.F. data in Table 12. By treating them in the way shown in Fig. 13, which is constructed on the same principle as Fig. 10, page 96,

I obtained the value for Fraternal Regression of $\frac{2}{3}$; that is to say, the unknown brother of a known man is probably only two-thirds as exceptional in Stature as he is. This is the same value as that obtained for the Regression from Mid-Parent to Son. However paradoxical the fact may seem at first, of there being such a thing as Fraternal Regression, a little reflection will show its reasonableness, which will become much clearer later on. In the meantime, we may recollect that the

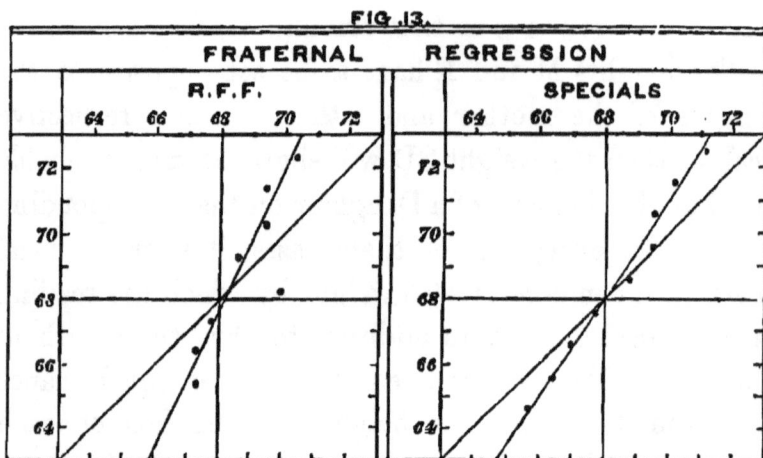

FIG. 13.

unknown brother has two different tendencies, the one to resemble the known man, and the other to resemble his race. The one tendency is to deviate from P as much as his brother, and the other tendency is not to deviate at all. The result is a compromise.

As the average Regression from either Parent to the Son is twice as great as that from a man to his Brother, a man is, generally speaking, only half as nearly related

to either of his Parents as he is to his Brother.   In
other words, the Parental kinship is only half as close
as the Fraternal.

We have now seen that there is Regression from the
Parent to his Son, from the Son to his Parent, and from
the Brother to his Brother.   As these are the only three
possible lines of kinship, namely, descending, ascending,
and collateral, it must be a universal rule that the unknown Kinsman, in any degree, of a known Man, is on
the average more mediocre than he.   Let $P\pm D$ be the
stature of the known man, and $P\pm D'$ the stature of his
as yet unknown kinsman, then it is safe to wager, in
the absence of all other knowledge, that D' is less
than D.

Squadron of Statures.—It is an axiom of statistics,
as I need hardly repeat, that every large sample taken
at random out of any still larger group, may be considered as identical in its composition, in such inquiries
as these in which we are now engaged, where minute
accuracy is not desired and where highly exceptional
cases are not regarded.   Suppose our larger group to
consist of a million, that is of 1000×1000 statures, and
that we had divided it at random into 1000 samples
each containing 1000 statures, and made Schemes of
each of them.   The 1000 Schemes would be practically
identical, and we might marshal them one behind the
other in successive ranks, and thereby form a "Squadron," numbering 1000 statures each way, and standing

upon a square base.   Our Squadron may be divided either into 1000 ranks or into 1000 files.   The ranks will form a series of 1000 identical Schemes, the files will form a series of 1000 rectangles, that are of the same breadth, but of dissimilar heights.   (See Fig. 14.)

It is easy by this illustration to give a general idea, to be developed as we proceed, of the way in which any large sample, A, of a Population gives rise to a group of Kinsmen, Z, so distant as to retain no family likeness

FIG.14.

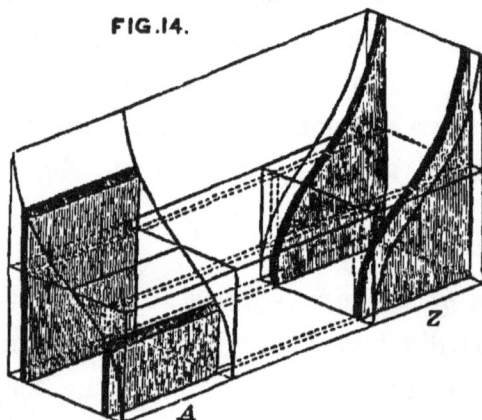

to A, but to be statistically undistinguishable from the Population generally, as regards the distribution of their statures.   In this case the samples A and Z would form similar Schemes.   I must suppose provisionally, for the purpose of easily arriving at an approximate theory, that tall, short, and mediocre Parents contribute equally to the next generation though this may not strictly be the case.[1]

---

[1] Oddly enough, the shortest couple on my list have the largest family, namely, sixteen children, of whom fourteen were measured.

Throw A into the form of a Squadron and not of a Scheme, and let us begin by confining our attention to the men who form any two of the rectangular files of A, that we please to select. Then let us trace their connections with their respective Kinsmen in Z. As the number of the Z Kinsmen to each of the A files is considered to be the same, and as their respective Stature-Schemes are supposed to be identical with that of the general Population, it follows that the two Schemes in Z derived from the two different rectangular files in A, will be identical with one another. Every other rectangular file in A will be similarly represented by another identical Scheme in Z. Therefore the 1,000 different rectangular files in A will produce 1,000 identical Schemes in Z, arranged as in Fig. 14.

Though all the Schemes in Z, contain the same number of measures, each will contain many more measures than were contained in the files of A, because the same kinsmen would usually be counted many times over. Thus a man may be counted as uncle to many nephews, and as nephew to many uncles. We will therefore (though it is hardly necessary to do so) suppose each of the files in Z to have been constructed from only a sample consisting of 1,000 persons, taken at random out of the more numerous measures to which it refers. By this treatment Z becomes an exact Squadron, consisting of 1,000 elements, both in rank and in file, and it is identical with A in its constitution, though not in its attitude. The ranks of Z, which are Schemes, have been derived from the files of A, which are rect-

angles, therefore the two Squadrons must stand at right
angles to one another, as in Fig. 14.   The upper surface
of A is curved in rank, and horizontal in file; that of
Z is curved in file and horizontal in rank.

The Kinsmen in nearer degrees than Z will be represented by Squadrons whose forms are intermediate
between A and Z.   Front views of these are shown in

FIG .I5.

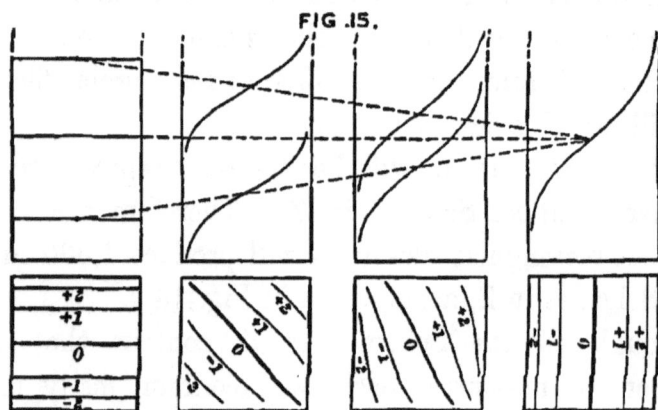

Fig. 15.   Consequently they will be somewhat curved
both in rank and in file.   Also as the Kinsmen of all
the members of a Population, in any degree, are themselves a Population having similar characteristics to
those of the Population of which they are part, it
follows that the elements of every intermediate Squadron
when they are broken up and sorted afresh into ordinary
Schemes, would form identical Schemes.   Therefore, it
is clear that a law exists that connects the curvatures in
rank and in file, of any Squadron.   Both of the curvatures are Curves of Distribution; let us call their
Q values respectively r and f.   Then if p be the Q of

I

the general Population, we arrive at a general equation
that is true for all degrees of Kinship; namely—

$$r^2 + f^2 = p^2 \qquad (1)$$

but r, the curvature in rank, is a regressed value of p,
and may be written wp, w being the value of the
Regression. Therefore the above equation may be put
in the form of

$$w^2 p^2 + f^2 = p^2 \qquad (2)$$

in which f is the Q of the Co-kinsmen in the given
degree.

It will be found that the intersection of the surfaces
of the Squadrons by a horizontal plane, whose height is
equal to P, forms in each case a line, whose general inclination to the ranks of A increases as the Kinship
becomes more remote, until it becomes a right angle in
Z. The progressive change of inclination is shown in
the small squares drawn at the base of Fig. 13, in which
the lines are the projections of contours drawn on the
upper surfaces of the Squadrons, to correspond with the
multiples there stated of values of p.

It will be understood from the front views of the
four different Squadrons, which form the upper part of
Fig. 13, how the Mid-Statures of the Kinsmen to the
Men in each of the files of A, gradually become more
mediocre in the successive stages of kinship until they
all reach absolute mediocrity in Z. This figure affords
an excellent diagramatic representation, true to scale,
of the action of the law of Regression in Descent. I
should like to have given in addition, a perspective
view of the Squadrons, but failed to draw them

clearly, after making many attempts. Their curvatures
are so delicate and peculiar that the eye can hardly
appreciate them even in a model, without turning it
about in different lights and aspects. A plaster model
of an intermediate form was exhibited at the Royal
Society by Mr. J. D. H. Dickson, when my paper on
Hereditary Stature was read, together with his solutions
of the problems that are given in the Appendix. I also
exhibited arrangements of files and ranks that were
made of pasteboard. Mr. Dixon mentioned that the
mathematical properties of a Surface of Frequency
showed that no strictly straight line could be drawn
upon it.

Successive Generations of a People.—We are far too
apt to regard common events as matters of course, that
require no explanation, whereas they may be problems
of much interest and of some difficulty, and still await
solution.

Why is it, when we compare two large groups of
persons selected at random from the same race, but
belonging to different generations, that they are usually
found to be closely alike? There may be some
small statistical dissimilarity due to well understood
differences in the general conditions of their lives, but
with this I am not concerned. The present question
is as to the origin of that statistical resemblance between
successive generations which is due to the strict processes of heredity, and which is commonly observed in
all forms of life.

In each generation, individuals are found to be tall and short, heavy and light, strong and weak, dark and pale; and the proportions of those who present these several characteristics in their various degrees, tend to be constant. The records of geological history afford striking evidences of this statistical similarity. Fossil remains of plants and animals may be dug out of strata at such different levels, that thousands of generations must have intervened between the periods at which they lived; yet in large samples of such fossils we may seek in vain for peculiarities that distinguish one generation from another, the different sizes, marks, and variations of every kind, occurring with equal frequency in all.

If any are inclined to reply that there is no wonder in the matter, because each individual tends to leave his like behind him, and therefore each generation must, as a matter of course, resemble the one preceding, the patent fact of Regression shows that they utterly misunderstand the case.

We have now reached a stage at which it has become possible to discuss the problem with some exactness, and I will do so by giving mathematical expression to what actually took place in the Statures of that sample of our Population whose life-histories are recorded in the R.F.F. data.

The Males and Females in Generation I. whose M has the value of P (viz., 68¼ inches), and whose Q is 1·7 inch, were found to group themselves as it were at random, into couples, and then to form a system of

Mid-Parents. This system had of course the same M as the general Population, but its Q was reduced to $\frac{1}{\sqrt{2}} \times 1\cdot7$ inch, or to $1\cdot2$ inch. It was next found when the Statures of the Mid-Parents, expressed in the form of $P + (\pm D)$, were sorted into groups in which D was the same (reckoning to the nearest inch), that a Co-fraternity sprang from each group, and that its M had the value of $P + (\pm \frac{2}{3}D)$. The system in which each element is a Mid-Co-Fraternity, must have the same M as before, of $68\frac{1}{4}$ inches, but its Q will be again reduced, namely from $1\cdot2$ inch to $\frac{2}{3} \times 1\cdot2$ inch, or to $0\cdot8$ inch. Lastly, the individual Co-Fraternals were seen to be dispersed from their respective Mid-Co-Fraternities, with a Q equal in each case to $1\cdot5$ inch. The sum of all of the Co-Fraternals forms the Population of Generation II. Consequently the members of Generation II. constitute a system that has an M of $68\frac{1}{4}$ inches and a Q equal to $\sqrt{[(0\cdot8)^2 + (1\cdot5)^2]}$, $= 1\cdot7$ inch. These values are identical with those in Generation I.; so the cause of their statistical similarity is tracked out.

There ought to be no misunderstanding as to the character of the evidence or of the reasoning upon which this analysis is based. A small but fair sample of the Population in two successive Generations has been taken, and its conditions as regards Stature have been strictly discussed. It was found that the distribution of Stature was sufficiently Normal to justify our ignoring any shortcomings in that respect. The transmutation

of female heights to their male equivalents was justified
by the fact that when the individual Statures of a group
of females are raised in the proportion of 100 to 108,
the Scheme drawn from them fairly coincides with that
drawn from male Statures. Marriage selection was found
to take no sufficient notice of Stature to be worth consideration ; neither was the number of children in
Fraternities found to be sensibly affected by the
Statures of their Parents. Again, it was seen to be
of no consequence when dealing statistically with the
offspring, whether their Parents were alike in stature or
not, the only datum deserving consideration being the
Stature of the Mid-Parent, that is to say, the average
value of (1) the Stature of the Father, and of (2) the
Transmuted Stature of the Mother. I fully grant that
not one of these deductions may be strictly exact, but
the error introduced into the conclusions by supposing
them to be correct proves not to be worth taking into
account in a first approximation.

Precisely the same may be said of the ulterior steps
in this analysis. Every one of them is based on the
properties of an ideally perfect curve, but in no case
has there been need to make any sensible departure
from the observed results, except in assigning a uniform
value to Q in the different Co-Fraternities. Strictly
speaking, that value was found to slightly rise or fall as
the Mid-Stature of the Co-Fraternity rose or fell. This
suggested the advisability of treating the whole inquiry
on the principle of the Geometric Mean, Appendix G.
I tried that principle in what seemed to be the most

hopeful case among my 18 schemes, but found the gain, if any, to be so small, that I did not care to go on with the experiment. It did not seem to deserve the additional trouble, and I was indisposed to do anything that was not really necessary, which might further confuse the reader. But had I possessed better data, I should have tried the Geometric Mean throughout. In doing so, every measure would be replaced by its logarithm, and these logarithms would be treated just as if they had been the observed values. The conclusions to which they might lead would then be re-transmuted to the numbers of which they were the logarithmic equivalents.

In short, we have dealt mathematically with an ideal population which has similar characteristics to those of a real population, and have seen how closely the behaviour of the ideal population corresponds in every stage to that of the real one. Therefore we have arrived at a closely approximate solution of the problem of statistical constancy, though numerous refinements have been neglected.

Natural Selection.—This hardly falls within the scope of our inquiry into Natural Inheritance, but it will be appropriate to consider briefly the way in which the action of Natural Selection may harmonise with that of pure heredity, and work together with it in such a manner as not to compromise the normal distribution of faculty. To do this, we must deal with the case that best represents the various possible

occurrences, namely that in which the mediocre members
of a population are those that are most nearly in
harmony with their circumstances. The harmony ought
to concern the aggregate of their faculties, combined
on the principle adopted in Table 3, after weighting
them in the order of their importance. We may deal
with any faculty separately, to serve as an example, if
its mediocre value happens to be that which is most
preservative of life under the majority of circumstances.
Such is Stature, in a rudely approximate degree, inasmuch as exceptionally tall or exceptionally short persons
have less chance of life than those of moderate size.

It will give more definiteness to the reasoning to
take a definite example, even though it be in part an
imaginary one. Suppose then, that we are considering
the stature of some animal that is liable to be hunted
by certain beasts of prey in a particular country. So
far as he is big of his kind, he would be better able
than the mediocrities to crush through thick grass and
foliage whenever he was scampering for his life, to jump
over obstacles, and possibly to run somewhat faster
than they. So far as he is small of his kind, he would
be better able to run through narrow openings, to
make quick turns, and to hide himself. Under the
general circumstances, it would be found that animals
of some particular stature had on the whole a better
chance of escape than any other, and if their race is
closely adapted to their circumstances in respect to
stature, the most favoured stature would be identical
with the M of the race. We already know that if we

call this value P, and write each stature under the
form of $P + x$ (in which x includes its sign), and if the
number of times with which any value $P + x$ occurs,
compared to the number of times in which P occurs,
be called y, then x and y are connected by the law
of Frequency of Error.

Though the impediments to flight are less unfavour-
able, on the average, to the stature P than to any other,
they will differ in different experiences.   The course of
one animal may chance to pass through denser foliage
than usual, or the obstacles in his way may be higher.
In that case an animal whose stature exceeded P would
have an advantage over mediocrities.   Conversely, the
circumstances might be more favourable to a small
animal.

Each particular line of escape would be most favourable to some particular stature, and whatever the value
of x might be, it is possible that the stature $P + x$
might in some cases be more favoured than any other.
But the accidents of foliage and soil in a country are
characteristic and persistent, and may fairly be considered as approximately of a typical kind.   Therefore
those that most favour the animals whose stature is
P will be more frequently met with than those that
favour any other stature $P + x$, and the frequency
of the latter occurrence will diminish rapidly as x
increases.   If the number of times with which any
particular value of $P + x$ is most favoured, as compared
with the number of times in which P is most favoured,
be called y', we may fairly assume that y' and x are

connected by the law of Frequency of Error. But
though the system of y values and that of y' values
may be both subject to the law, it is not for a moment
to be supposed that their Q values are necessarily
the same.

We have now to show how a large population of
animals becomes reduced by the action of natural
selection to a smaller one, in which the M value of the
statures is unchanged, while the Q value is decreased.

To do this we must first consider the population to
have grown up entirely shielded from causes of premature mortality; call their number N. Then suppose
them to be assailed by all the lethal influences that have
no reference to stature. These would reduce their
number to N', but by the hypothesis, the values of
M and of Q would remain unaffected. Next let the
influences that act selectively on stature, further reduce
the numbers to S; these being the final survivors.
We have seen that :—

$y=$ the number of individuals who have the stature
$P\pm x$, counting those who have the stature P, as 1.

$y'=$ the number of times in which $P\pm x$ is the most
favoured stature, counting those in which P is the
most favoured, as 1.

Then $yy'=$ the number of times that individuals of
the stature $P\pm x$ are selected, counting those in which
individuals of the stature P are selected, as 1.

As the relation between y and x, and between y' and
x are severally governed by the law of Frequency of
Error, it follows directly from the formula by which

that law is expressed, that the relation between yy' and x is also governed by it. The value of P of course remains the same throughout, but the Q in the system of yy' values is necessarily less than that in the system of y values.

It might well be that natural selection would favour the indefinite increase of numerous separate faculties, if their improvement could be effected without detriment to the rest ; then, mediocrity in that faculty would not be the safest condition. Thus an increase of fleetness would be a clear gain to an animal liable to be hunted by beasts of prey, if no other useful faculty was thereby diminished.

But a too free use of this "if" would show a jaunty disregard of a real difficulty. Organisms are so knit together that change in one direction involves change in many others ; these may not attract attention, but they are none the less existent. Organisms are like ships of war constructed for a particular purpose in warfare, as cruisers, line of battle ships, &c., on the principle of obtaining the utmost efficiency for their special purpose. The result is a compromise between a variety of conflicting desiderata, such as cost, speed, accommodation, stability, weight of guns, thickness of armour, quick steering power, and so on. It is hardly possible in a ship of any long established type to make an improvement in any one of these respects, without a sacrifice in other directions. If the fleetness is increased, the engines must be larger, and more space must be given up to coal, and this diminishes the remaining accommodation. Evolution may produce an altogether new type of vessel that shall be more efficient than the old one, but when a particular type of vessel has become adapted to its functions through long experience it is not possible to produce a mere variety of its type that shall have increased efficiency in some one particular without detriment to the rest. So it is with animals.

Variability in Fraternities.—Human Fraternities are far too small to admit of their Q being satisfactorily measured by the direct method. We are obliged to have recourse to indirect methods, of which no less than four are available. I shall apply each of them to both the Special and to the R.F.F. data; this will give 8 separate estimates of its value, which in the meantime will be called *b*. The four methods are as follow:

First method; by Fraternities each containing the same number of persons :—Let me begin by saying that I had already found in the large Fraternities of Sweet Peas, that the sizes of individuals of whom they consisted were normally distributed, and that their Q was independent of the size, or of the Stature as we may phrase it, of the Mid-Fraternity. We have also seen that the Q is practically the same in all Co-fraternities of men. Therefore it is reasonable to expect that it will also be found to be the same in all their Fraternities, though owing to their small size we cannot assure ourselves of the fact by direct evidence. We will assume this to be the case for the present; it will be seen that the results justify the assumption.

The value of the M of a small Fraternity may be much affected by the addition or subtraction even of a single member, it may therefore be called the *apparent* M, to be distinguished from the *true* M, from which its members would be found to be dispersed, if there had been many more of them.   The apparent M approximates towards the true M as the Fraternity increases in size, though at a much slower rate.   We have now somehow to get at this true M.   For distinction and for brevity let us call the *apparent* M of any small Fraternity (MF′), and that of the corresponding *true* M (MF).   Then (MF) may be deduced from (MF′) as follows :—

We will begin by allowing ourselves for the moment to imagine the existence of an exceedingly large Fraternity, far more numerous than is physiologically possible, and to suppose that its members vary among themselves just as widely, neither more nor less so, than in the small Fraternities of real life.   The (MF′) of our large ideal Fraternity will therefore be identical with its (MF), and its Q will be the same as b.   Next, take at random out of this huge ideal Fraternity a large number of small samples, each consisting of the same number, n, of brothers, and call the apparent Mid-values in the several samples, (MF'_1), (MF'_2), &c.   It can easily be shown that (MF'_1), (MF'_2), &c., will be so distributed about the common centre of (MF), that the Prob. Deviation of any one of them from it, that is to say, the Q of their system will $= b \times \frac{1}{\sqrt{n}}$.   If $n = 1$, then the Prob. Deviation becomes b, as it should.   If $n = 2$, the Prob.

Deviation is determined by the same problem as that which concerned the Q of the Mid-Parentages (page 87), where it was shown to be $b \times \frac{n}{\sqrt{2}}$. By similar reasoning, when $n = 3$, the Prob. Deviation becomes $b \times \frac{1}{\sqrt{3}}$, and so on. When n is infinitely large, the Prob. Deviation is 0; that is to say, the (MF') values do not differ at all from their common (MF).

Now if we make a collection of human Fraternities, each consisting of the same number, n, of brothers, and note the differences between the (MF') in each fraternity and the individual brothers, we shall obtain a system of values. By drawing a Scheme from these in the usual way, we are able to find their Q; let us call it d. We then determine b in terms of d, as follows :—The (MF') values are distributed about their common (MF), each with the Prob. Deviation of $b \times \frac{1}{\sqrt{n}}$, and the Statures of the individual Brothers are distributed about their respective (MF') values, each with the Prob. Deviation d. The compound result is the same as if the statures of the individual brothers had been distributed about the common (MF), each with the Prob. Deviation b,

$$\text{consequently } b^2 = d^2 + \frac{b^2}{n}, \text{ or } b^2 = \frac{n}{n-1} d^2.$$

I determined d by observation for four different values of n, having taken four groups of Fraternities, consisting respectively of 4, 5, 6, and 7 brothers, as shown in Table 14. Substituting these four observed values in turns for d in the above formula, I obtained four independent values of b, which are respectively 1·01, 1·01, 1·20, and 1·08 ; the mean of these is 1·07.

Second Method; from the mean value of Fraternal Regression :—We may look on the Population as composed of a system of Fraternities.   Call their respective true centres (see last paragraph) (MF_1), (MF_2), &c. These will be distributed about P with an as yet unknown Prob. Deviation, that we will call c.   The individual members of each Fraternity will of course be distributed from their own (MF) with a Q equal to b.
Then $(1·7)^2 = c^2 + b^2$ (1)
Let $P + (\pm F_n)$ be the stature of any individual, and let $P + (\pm M\,F_n)$ be that of the **M** of his Fraternity, then Problem 4 (page 69) shows us that :—

the *most probable* value of $\dfrac{(MF_n)}{F_n}$ is $\sqrt{\dfrac{c^2}{b^2 + c^2}}$ (2)

This is also the value of Fraternal Regression, and therefore equal to $\tfrac{2}{3}$.   Substituting in (2), and replacing c by the value given by (1), we obtain $b = 0·98$ inch.

Third Method; by the Variability in the value of individual cases of Fraternal Regression :—The figures in each line of Table 13 are found to have a Q equal to 1·24 inch, and they are the results of two independent systems of variation. First, the several (MF) values (see last paragraph) are dispersed from the **M** of all of them with a Q that we will call v.   Secondly the

individual brothers in each Fraternity are dispersed
from their own (MF) with a Q equal to b.

Hence $(1 \cdot 24)^2 = v^2 + b^2$.

But it is shown Problem 5 that $v = \dfrac{bc}{\sqrt{(b^2 + c^2)}}$;

therefore $(1 \cdot 24)^2 = b^2 + \dfrac{b^2 c^2}{b^2 + c^2}$.

Substituting for c^2 its value of $(1 \cdot 7)^2 - b^2$ (see last paragraph), we obtain $b = 0 \cdot 98$ inch.

Fourth Method; from differences between pairs of
brothers taken at random :—In the fourth method,
Pairs of Brothers are taken at random, and the Differences between the statures in each pair are noted ; then,
under the following reservation, any one of these
differences would have the Prob. value of $\sqrt{2} \times b$. The
reservation is, that only as many Differences should be
taken out of each Fraternity as are independent. A
Fraternity of n brothers admits of $\frac{n(n-1)}{2}$ possible pairs,
and the same number of Differences ; but as no more
than $n - 1$ of these are independent, that number only
of the Differences should be taken. I did not appreciate this necessity at first, and selected pairs of brothers
on an arbitrary system, which had at all events the
merit of not taking more than four sets of Differences
from any one Fraternity however large it might be.
It was faulty in taking three Differences instead of only
two, out of a Fraternity of three brothers, and four
Differences, instead of only three, from a Fraternity of

four brothers, and therefore giving an increased weight to those Fraternities, but in other respects the system was hardly objectionable. The introduced error must be so slight as to make it scarcely worth while now to go over the work again. By the system adopted, I found the Prob. Difference to be 1·55, which divided by $\sqrt{2}$ gives $b = 1·10$ inch.

Thus far we have dealt with the special data only. The less trustworthy R.F.F. give larger values of b in every case. An epitome of all the results appears in the following table :—

Methods and data.	Values of b obtained by different methods and from different data.	
	From Special Data.	From R.F.F. data.[1]
(1) From Fraternities each containing the same number of persons	1·07	1·38
(2) From the mean value of Fraternal Regression............	0·98	1·31
(3) From the Variability of Fraternal Regression......... ..	1·10	1·14
(4) From Pairs of Brothers taken at random.................	1·10	1·35
Mean.............,.....	1·06	

The data used in the four methods are somewhat different. In (1) I could not deal with small Fraterni-

---

[1] The R.F.F. results were obtained from brothers only and not from transmuted sisters, except in method (2), where the paucity of the data compelled me to include them.

ties, so all were disregarded that contained fewer than
four individuals.   In (2) and (3) I could not with
safety use large Fraternities.   In (4) the method of
selection was, as we have seen, quite indifferent.   This
makes the accordance of the results derived from the
Special data all the more gratifying.   Those from the
R.F.F. data accord less well together.   The R.F.F.
measures are not sufficiently exact for use in these
delicate calculations.   Their results, being compounded
of b and of their tendency to deviate from exactness,
are necessarily too high, and should be discarded.   I
gather from all this that we may safely consider the
value of b to be less than 1·06, and that allowing for
some want of precision in the Special data, the very
convenient value of 1·00 inch may reasonably be
adopted.

Trustworthiness of the Constants.—There is difficulty
in correcting the results obtained from the R.F.F. data,
though we can make some estimate of their general
inaccuracy as compared with the Special data.   The
reason of the difficulty is that the inaccuracy cannot
be ascribed to an uncertainty of equal ± amount in
every entry, such as might be due to a doubt of
" shoes off " or " shoes on."   If it were so, the Prob.
Error of a single value of the R.F.F. would be greater
than that of one of the Specials, whereas it proves to
be the same.   It is likely that the inaccuracy is a compound first of the uncertainty above mentioned, whose
effect would be to increase the value of the Prob. Error,

and secondly of a tendency on the part of my correspondents to record medium statures when they were
in doubt, whose effect would be to reduce the value of
the Prob. Error.   The R.F.F. data in Table 12 run so
irregularly that I cannot interpret them with any
assurance.   The value they give for Fraternal Regression
certainly does not exceed $\frac{1}{2}$, and therefore a correction,
amounting to no less than $\frac{1}{3}$ of its amount, is required
to bring it to a parity with that derived from the
Special data (because $\frac{1}{2} + \frac{1}{3} \times \frac{1}{2} = \frac{2}{3}$).   Hence it
might be argued, that the value of Regression from
Mid-Parent to Son, which the R.F.F. data gave as $\frac{2}{3}$,
ought to receive a similar correction.   If so, it would
be raised to $\frac{2}{3} + \frac{2}{9} = \frac{8}{9}$; but I cannot believe this
high value to be correct.   My first estimate made
from the R.F.F. data, was $\frac{3}{5}$, as already mentioned.   If
this be adopted, the corrected value would be $\frac{4}{5}$, or $\frac{8}{10}$
instead of $\frac{8}{9}$, which might possibly pass.   Curiously
enough, this value of $\frac{4}{5}$ for Regression from Mid-Parent
to Son, coincides with the value of $\frac{2}{5}$ for Regression
from a single Parent to Son, which the direct observations showed (see page 99), but which owing to their
paucity and to the irregularity of the way in which
they ran, I rejected and still reject, at least for the
present.   While sincerely desirous of obtaining a
revised value of average Filial Regression from entirely
different and more accurate groups of data, the provisional value already adopted of $\frac{2}{3}$ from Mid-Parent
to Son may be accepted as being near enough for the
present.   It is impossible to revise one datum in the

R.F.F. series without revising all, as they hang together
and support one another.

General View of Kinship.—We are now able to deal
with the distribution of statures among the Kinsmen in
every near degree, of persons whose statures we know,
but whose ancestral statures we either do not know, or
do not care to take into account.   We are able to calculate Tables for every near degree of Kinship on the form
of Table 11, and to reconstruct that same Table in a
shape free from irregularities.   We must first find the
Regression, which we may call w, appropriate to the
degree of Kinship in question.   Then we calculate a
value f for each line of a Table corresponding in form to
that of Table 11, in which f was found to be equal to
1·50 inch.   We deduce the value of f from that of w by
means of the general equation $p^2w^2 + f^2 = p^2$, p being
equal to 1·7 inch.   The values to be inserted in the
several lines are then calculated from the ordinary table
(Table 5) of the " probability integral."

As an example of the first part of the process, let us
suppose we are about to construct a table of Uncles and
their Nephews, we find w and f as follows : A Nephew
is the son of a Brother, therefore in this case we have
$w = \frac{1}{3} \times \frac{2}{3} = \frac{2}{9}$; whence $f = 1\cdot66$.

The Regression, which we call w, is a convenient and
correct measure of family likeness.   If the resemblance
of the Kinsman to the Man, was on the average as
perfect as that of the Man to his own Self, there would
be no Regression at all, and the value of w would be 1.

TABLE OF DATA FOR CALCULATING TABLES OF DISTRIBUTION OF STATURE AMONG THE KINSMEN OF PERSONS WHOSE STATURE IS KNOWN.

From group of persons of the same Stature, to their Kinsmen in various near degrees.	Mean regression=w.	$Q = f$ $= p \times \sqrt{(1-w^2)}.$
Mid-parents to Sons............	2 / 3	1·27
Brothers to Brothers	2 / 3	1·27
Fathers or Sons to } Sons or Fathers }	1 / 3	1·60
Uncles or Nephews to } Nephews or Uncles }	2 / 9	1·66
Grandsons to Grandparents...	1 / 9	{ Practically that of Population, or 1·7 inch.
Cousins to Cousins	2 / 27	

On the other hand, if the Kinsmen were on the average no more like the Man than if they had been a group picked at random out of the general Population, then the Regression to P would be complete. The M of the Kinsmen, which is expressed by $P + w(\pm D)$, would in that case become P, whatever might have been the value of D; therefore w must $= 0$. We see by the preceding Table that as a general rule, Fathers or Sons should be held to be only one-half as near in blood as Brothers, and Uncles and Nephews to be one-third as near in blood as Brothers. Cousins are $4\frac{1}{2}$ times as remote as Fathers or as Sons, and 9 times as remote as Brothers. I do not extend the table further, because considerations would have to be taken into account that will be discussed in the next Section.

The remarks made in a previous chapter about

heritages that blend and those that are mutually exclusive, must be here borne in mind. It would be a poor
prerogative to inherit say the fifth part of the peculiarity
of some gifted ancestor, but the chance of 1 to 5, of
inheriting the whole of it, would be deservedly prized.

Separate Contribution of each Ancestor.—In making
the statement that Mid-Parents whose Stature is $P \pm D$
have children whose average stature is $P \pm \frac{2}{3}D$, it is
supposed that no separate account has been taken of
the previous ancestry. Yet though nothing may be
known of them, something is tacitly implied and has
been tacitly allowed for, and this requires to be eliminated before we can learn the amount of the Parental
bequest, pure and simple. What that something is, we
must now try to discover. When speaking of converse
Regression, it was shown that a peculiarity in a Man
implied a peculiarity of $\frac{1}{3}$ of that amount in his Mid-
Parent. Call the peculiarity of the Mid-Parent D, then
the implied peculiarity of the Mid-Parent of the Mid-
Parent, that is of the Mid-Grand-Parent of the Man,
would on the above supposition be $\frac{1}{3}D$, that of the Mid-
Great-Grand-Parent would be $\frac{1}{9}D$, and so on. Hence
the total bequeathable property would amount to
$D(1 + \frac{1}{3} + \frac{1}{9} + \&c.) = D\frac{3}{2}$.

Do the bequests from each of the successive generations reach the child without any, or what, diminution
by the way? I have not sufficient data to yield a direct
reply, and must therefore try limiting suppositions.

First, suppose the bequests by the various generations

to be equally taxed; then, as an accumulation of ancestral contributions whose sum amounts to $D\frac{3}{2}$ yields an
effective heritage of only $D\frac{2}{3}$, it follows that each piece
of heritable property must have been reduced to $\frac{4}{9}$ of its
original amount, because $\frac{3}{2} \times \frac{4}{9} = \frac{2}{3}$.

Secondly, suppose the tax not to be uniform, but to
be repeated at each successive transmission, and to be
equal to $\dfrac{1}{r}$ of the amount of the property at each
stage. In this case the effective heritage would be
$$D\left(\frac{1}{r} + \frac{1}{3r^2} + \frac{1}{3^2 r^3} + -\right) = D\,\frac{3}{3r-1}\ ,$$ which must, as
before, be equal to $D\frac{2}{3}$; whence $\dfrac{1}{r} = \dfrac{6}{11}$

Thirdly, it might possibly be supposed that the Mid-
Ancestor in a remote generation should on the average
contribute more to the child than the Mid-Parent, but
this is quite contrary to what is observed. The descendants of what was "pedigree wheat," after being left to
themselves for many generations, show little or no trace
of the remarkable size of their Mid-Ancestors in the
generations just before they were left to themselves,
though the offspring of those Mid-Ancestors in the first
generation did so unmistakably.

The results of our only two valid limiting suppositions
are therefore, (1) that the Mid-Parental peculiarities,
pure and simple, influence the offspring to $\frac{4}{9}$ of their
amount; (2) that they influence it to $\frac{6}{11}$ of their amount.
These values differ but slightly from $\frac{1}{2}$, and their mean
is closely $\frac{1}{2}$, so we may fairly accept that result. Hence

the influence, pure and simple, of the Mid-Parent may be taken as $\frac{1}{2}$, and that of the Mid-Grand-Parent as $\frac{1}{4}$, and so on. Consequently the influence of the individual Parent would be $\frac{1}{4}$, and of the individual Grand-Parent $\frac{1}{16}$, and so on. It would, however, be hazardous on the present slender basis, to extend this sequence with confidence to more distant generations.

Pedigree Moths.—I am endeavouring at this moment to obtain data that will enable me to go further, by breeding Pedigree Moths, thanks to the aid of Mr. Frederick Merrifield. The moths *Selenia Illustraria* and *Illunaria* are chosen for the purpose, partly on account of their being what is called double brooded; that is to say, they pass normally through two generations in a single year, which is a great saving of time to the experimenter. They are hardy, prolific and variable, and are found to stand chloroform well, previously to being measured and then paired. Every member of each Fraternity is preserved along three lines of descent—one race of long-winged moths, one of medium-winged, and one of short-winged moths. The three parallel sets are reared under identical conditions, so that the medium series supplies a trustworthy relative base, from which to measure the increasing divergency of the others. No one can be sure of the success of any extensive breeding experiment, but this attempt has been well started and seems to present no peculiar difficulty. Among other reasons for choosing moths for the purpose, is that they are born adults, not changing in stature after they have emerged from the chrysalis and shaken out their wings. Their families are of a convenient size for statistical purposes, say from 50 to 100, neither too few to make satisfactory Schemes, nor unmanageably large. They can be mounted as we all know, after their death, with great facility, and be remeasured at leisure. An intelligent and experienced person can carry on a large breeding establishment in a small room, supplemented by a small garden. The methods used and the results up to last spring, have been described by Mr. Merrifield in papers read February and December 1887, and printed in the Transactions of the Entomological Society. I speak of this now, in hopes of attracting the attention of some who are competent and willing to carry on collateral experiments with the same breed, or with altogether different species of moths.

CHAPTER VIII.

Preliminary Remarks.—In this chapter I will test the conclusions respecting stature by an examination into hereditary Eye-colour. Supposing all female measures to have been transmuted to their male equivalents, it has been shown (1) that the possession of each unit of *peculiarity* of stature in a man [that is of each unit of difference from the average of his race] when the man's ancestry is unknown, implies the existence on an average of just one-third of a unit of that peculiarity in his "Mid-Parent," and consequently of the same amount in each of his parents; also just one-third of a unit in his Son; (2) that each unit of peculiarity in each ancestor taken singly, is reduced in transmission according to the following average scale ;— a Parent transmits only $\frac{1}{4}$, and a Grand-Parent only $\frac{1}{16}$.

Stature and Eye-colour are not only different as qualities, but they are more contrasted in hereditary behaviour than perhaps any other common qualities. Parents of different Statures usually transmit a blended heritage to their children, but parents of different Eye-colours usually transmit an alternative heritage. If one parent is as much taller than the average of his or her sex as the other parent is shorter, the Statures of their children will be distributed, as we have already seen, in nearly the same way as if the parents had both been of medium height. But if one parent has a light Eye-colour and the other a dark Eye-colour, some of the children will, as a rule, be light and the rest dark; they will seldom be medium eye-coloured, like the children of medium eye-coloured parents. The blending in Stature is due to its being the aggregate of the quasi-independent inheritances of many separate parts, while Eye-colour appears to be much less various in its origin. If notwithstanding this two-fold difference between the qualities of Stature and Eye-colour, the shares of hereditary contribution from the various ancestors are alike in the two cases, as I shall show that they are, we may with some confidence expect that the law by which those hereditary contributions are found to be governed, may be widely, and perhaps universally applicable.

Data.—My data for hereditary Eye-colour are drawn from the same collection of "Records of Family Faculties" ("R.F.F.") as those upon which the inquiries into hereditary Stature were principally based. I have analysed the general value of these data in respect to

Stature, and shown that they were fairly trustworthy. I think they are somewhat more accurate in respect to Eye-colour, upon which family portraits have often furnished direct information, while indirect information has been in other cases obtained from locks of hair that were preserved in the family as mementos.

Persistence of Eye-colour in the Population.—The first subject of our inquiry must be into the existence of any slow change in the statistics of Eye-colour in the English population, or rather in that particular part of it to which my returns apply, that ought to be taken into account before drawing hereditary conclusions. For this purpose I sorted the data, not according to the year of birth, but according to generations, as that method best accorded with the particular form in which all my R.F.F. data are compiled. Those persons who ranked in the Family Records as the "children" of the pedigree, were counted as generation I.; their parents, uncles and aunts, as generation II.; their grandparents, great uncles, and great aunts, as generation III.; their great grandparents, and so forth, as generation IV. No account was taken of the year of birth of the "children," except to learn their age; consequently there is much overlapping of dates in successive generations. We may however safely say, that the persons in generation I. belong to quite a different period to those in generation III., and the persons in II. to those in IV. I had intended to exclude all children under the age of eight years, but in this particular branch of the inquiry, I fear that some cases of young children have been accidentally included.   I would willingly have taken a later limit than eight years, but could not spare the data that would in that case have been lost to me.

A great variety of terms are used by the various compilers of the "Family Records" to express Eye-colours.   I began by classifying them under the following eight heads ;—1, light blue ; 2, blue, dark blue ; 3, grey, blue-green ; 4, dark grey, hazel ; 5, light brown ; 6, brown ; 7, dark brown ; 8, black.   Then I constructed Table 15.

The diagram, page 143, clearly conveys the significance of the figures in Table 15.   Considering that the groups into which the observations are divided are eight in number, the observations are far from being sufficiently numerous to justify us in expecting clean results ; nevertheless the curves come out surprisingly well, and in accordance with one another.   There can be little doubt that the change, if any, during four successive generations is very small, and much smaller than mere memory is competent to take note of.   I therefore disregard a current popular belief in the existence of a gradual darkening of the British population, and shall treat the eye-colours of those classes of our race who have contributed the records, as having been statistically persistent during the period under discussion.

The concurrence of the four curves for the four several generations, affords internal evidence of the trustworthiness of the data.   For supposing we had curves that exactly represented the true Eye-colours for the four generations, they would either be concurrent or they would not. If these curves were concurrent, the errors in the R.F.F. data must have been so curiously distributed as to preserve the concurrence. If these curves were not concurrent, then the errors in the R.F.F. data must have been so curiously distributed as to neutralise the non-concurrence. Both of these suppositions are improbable, and we must conclude that the curves really agree, and that the R.F.F. errors are not large enough to spoil the agreement. The close similarity of the two curves, derived respectively from the whole of the male and the whole of the female data, and the more perfect form of the curve derived from the aggregate of all the cases, are additional evidences in favour of the goodness of the data on the whole.

Fundamental Eye-colours.—It is agreed among writers (*cf.* A. de Candolle, see footnote overleaf) that the one important division of eye-colours is into the light and the dark. The medium tints are not numerous, but may be derived from any one of four distinct origins. They may be hereditary with no notable variation, they may be varieties of light parentage, they may be varieties of dark parentage, or they may be blends. Medium tints are classed in my list under the heading " 4. Dark grey, hazel; " these form only 12·7 per cent. of all the observed cases. In medium tints, the outer portion of the iris is often of a dark grey colour,

Percentages of the Various Eye-colours in Four Successive Generations.

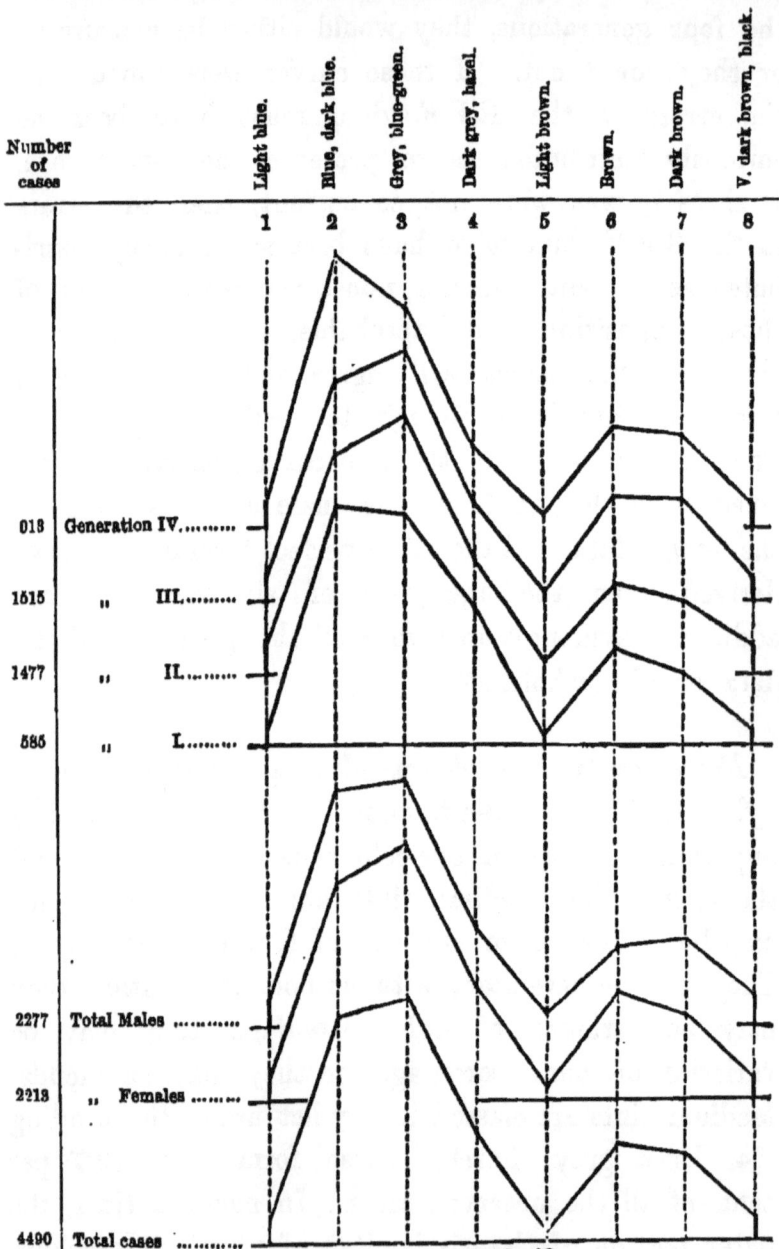

FIG. 16.

and the inner of a hazel. The proportion between the grey and the hazel varies in different cases, and the eye-colour is then described as dark grey or as hazel, according to the colour that happens most to arrest the attention of the observer. For brevity, I will henceforth call all intermediate tints by the one name of hazel.

I will now investigate the history of those hazel eyes that are variations from light or from dark respectively, or that are blends between them. It is reasonable to suppose that the residue which were inherited from hazel-eyed parents, arose in them or in their predecessors either as variations or as blends, and therefore the result of the investigation will enable us to assort the small but troublesome group of hazel eyes in an equitable proportion between light and dark, and thus to simplify our inquiry.

The family records include 168 families of brothers and sisters, counting only those who were above eight years of age, in whom one member at least had hazel eyes. For distinction I will describe these as "hazel-eyed families;" not meaning thereby that all the children have that peculiarity, but only one or more of them. The total number of the brothers and sisters in the 168 hazel-eyed families is 948, of whom 302 or about one-third have hazel eyes. The eye-colours of all the 2 × 168, or 336 parents, are given in the records, but only those of 449 of the grandparents, whose number would be 672, were it not for a few cases of cousin marriages. Thus I have information concerning about only two-thirds of the grandparents, but this will suffice for our purpose. The results are given in Table 16.

It will be observed that the distribution of eye-colour among the grandparents of the hazel-eyed families is nearly identical with that among the population at large. But among the parents there is a notable difference; they have a decidedly larger percentage of light eye-colour and a slightly smaller proportion of dark, while the hazel element is nearly doubled. A similar change is superadded in the children. The total result in passing from generations III. to I., is that the percentage of the light eyes is diminished from 60 or 61 to 45, therefore by one quarter of its original amount, and that the percentage of the dark eyes is diminished from 26 or 27 to 23, that is by about one-eighth of its original amount, the hazel element in either case absorbing the difference. It follows that the chance of a light-eyed parent having hazel offspring, is about twice as great as that of a dark-eyed parent. Consequently, since hazel is twice as likely to be met with in any given light-eyed family as in a given dark-eyed one, we may look upon two-thirds of the hazel eyes as being fundamentally light, and one-third of them as fundamentally dark. I shall allot them rateably in that proportion between light and dark, as nearly as may be without using fractions, and so get rid of them. M. Alphonse de Candolle[1] has

[1] Hérédité de la Couleur des Yeux dans l'Espèce humaine," par M. Alphonse de Candolle. "Arch. Sc. Phys. et Nat. Geneva," Aug. 1884, 3rd period. vol. xii. p. 97.

also shown from his data, that *yeux gris* (which I take
to be the equivalent of my *hazel*) are referable to a
light ancestry rather than to a dark one, but his data
are numerically insufficient to warrant a precise estimate
of the relative frequency of their derivation from each
of these two sources.

In the following discussion I shall deal only with
those fraternities in which the Eye-colours are known
of the two Parents and of the four Grand-Parents.
There are altogether 211 of such groups, containing
an aggregate of 1023 children. They do not, however,
belong to 211 different family stocks, because each
stock which is complete up to the great grand-parents
inclusive (and I have fourteen of these) is capable
of yielding three such groups. Thus, group 1 contains
a, the "children;" *b*, the parents; *c*, the grandparents. Group 2 contains *a*, the father of the
"children" and his brothers and his sisters; *b*, the
parents of the father; *c*, the grand-parents of the
father. Group 3 contains the corresponding selections
on the mother's side. Other family stocks furnish two
groups. Out of these and other data, Tables 19 and
20 have been made. In Table 19 ·I have grouped
the families together whose two parents and four grandparents present the same combination of Eye-colour,
no group, however, being accepted that contains less
than twenty children. The data in this table enable
us to test the *average* correctness of the law I desire
to verify, because many persons and many families
appear in the same group, and individual peculiarities

tend to neutralise each other. In Table 20 I have separately classified on the same system all the families, 78 in number, that consist of six or more children. These data enable us to test the trustworthiness of the law as applied to *individual* families. It will be seen from my way of discussing them, that smaller fraternities than these could not be advantageously dealt with.

It will be noticed that I have not printed the number of dark-eyed children in either of these tables. They are implicitly given, and are instantly to be found by subtracting the number of light-eyed children from the total number of children. Nothing would have been gained by their insertion, while compactness would have been sacrificed.

The entries in the tables are classified, as I said, according to the various combinations of light, hazel, and dark Eye-colours in the Parents and Grand-Parents. There are six different possible combinations among the two Parents, and 15 among the four Grand-Parents, making 6 × 15, or 90 possible combinations altogether. The number of observations are of course by no means evenly distributed among the classes. I have no returns at all under more than half of them, while the entries of two light-eyed Parents and four light-eyed Grand-Parents are proportionately very numerous.

The question of marriage selection in respect to Eye-colour, has been already discussed briefly in p. 86. It is a less simple statistical question than at a first sight it may appear to be, so I will not discuss it farther.

Principles of Calculation.—I have next to show how the expectation of Eye-colour among the children of a given family is to be reckoned on the basis of the same law that held in respect to stature, so that calculations of the probable distribution of Eye-colours may be made. They are those that fill the three last columns of Tables 19 and 20, which are headed I., II., and III., and are placed in juxtaposition with the observed facts entered in the column headed "Observed." These three columns contain calculations based on data limited in three different ways, in order the more thoroughly to test the applicability of the law that it is desired to verify. Column I. contains calculations based on a knowledge of the Eye-colours of the Parents only; II. contains those based on a knowledge of those of the Grand-Parents only; III. contains those based on a knowledge of those both of the Parents and of the Grand-Parents, and of them only.

I. Eye-colours given of the two Parents—

Let the letter S be used as a symbol to signify the subject (or person) for whom the expected heritage is to be calculated. Let F stand for the words "a parent of S;" G_1 for "a grandparent of S;" G_2 for "a great-grandparent of S," and so on.

We must begin by stating the problem as it would stand if Stature was under consideration, and then modify it so as to apply to Eye-colour. Suppose then, that the amount of the peculiarity of Stature possessed by F is equal to D, and that nothing whatever

is known with certainty of any of the ancestors of
S except F. We have seen that though nothing may
actually be known, yet that something definite is implied
about the ancestors of F, namely, that each of his two
parents (who will stand in the order of relationship
of G_1 to S) will on the average possess $\frac{1}{3}$D. Similarly
that each of the four grandparents of F (who will stand
in the order of G_2 to S) will on the average possess
$\frac{1}{9}$D, and so on. Again we have seen that F, on the
average, transmits to S only $\frac{1}{4}$ of his peculiarity; that
G_1 transmits only $\frac{1}{16}$; G_2 only $\frac{1}{64}$, and so on. Hence
the aggregate of the heritages that may be expected
to converge through F upon S, is contained in the
following series :—

$$D\left\{\frac{1}{4}+2\left(\frac{1}{3}\times\frac{1}{2^4}\right)+4\left(\frac{1}{9}+\frac{1}{2^6}\right)+ \text{ \&c. }\right\}$$
$$=D\left\{\frac{1}{2^2}+\frac{1}{2^3.3}+\frac{1}{2^4.3^2}+ \text{ \&c. }\right\}=D\times0\cdot30.$$

That is to say, each parent must in this case be
considered as contributing 0·30 to the heritage of the
child, or the two parents together as contributing 0·60,
leaving an indeterminate residue of 0·40 due to the
influence of ancestry about whom nothing is either
known or implied, except that they may be taken as
members of the same race as S.

In applying this problem to Eye-colour, we must bear
in mind that the fractional chance that each member
of a family will inherit either a light or a dark Eye-
colour, must be taken to mean that that same fraction

of the total number of children in the family will
probably possess it.   Also, as a consequence of this
view of the meaning of a fractional chance, it follows
that the residue of 0·40 must be rateably assigned
between light and dark Eye-colour, in the proportion
in which those Eye-colours are found in the race
generally, and this was seen to be (see Table 16) as
61·2 : 26·1 ; so I allot 0·28 out of the above residue
of 0·40 to the heritage of light, and 0·12 to the heritage
of dark.   When the parent is hazel-eyed I allot $\frac{2}{3}$ of
his total contribution of 0·30, *i.e.*, 0·20 to light, and
$\frac{1}{3}$, *i.e.* 0·10 to dark.   These chances are entered in the
first pair of columns headed I. in Table 17.

The pair of columns headed I. in Table 18 shows
the way of summing the chances that are given in the
columns that have a similar heading in Table 17.   By
the method there shown, I calculated all the entries
that appear in the columns with the heading I. in Tables
19 and 20.

II. Eye-colours given of the four Grand Parents—

Suppose D to be possessed by G_1 and that nothing
whatever is known with certainty of any other ancestor
of S.   Then it has been shown that the child of G_1
(that is F) will possess $\frac{1}{3}D$; that each of the two parents
of G_1 (who stand in the relation of G_2 to S) will also
possess $\frac{1}{3}D$; that each of the four grandparents of G_1
(who stand in the relation of G_3 to S) will possess $\frac{1}{3}D$,
and so on.   Also it has been shown that the shares
of their several peculiarities that will on the average
be transmitted by F, G_1, G_2, &c., are $\frac{1}{4}$, $\frac{1}{16}$, $\frac{1}{64}$, &c.,

respectively. Hence the aggregate of the probable heritages from G_1 are expressed by the following series :—

$$D\left\{\frac{1}{3}\times\frac{1}{2^2}+1\times\frac{1}{2^4}+\frac{1}{3}\times 2\times\frac{1}{2^6}+\frac{1}{9}\times 4\times\frac{1}{2^8}+\&c.\right\}$$

$$= D\left\{\frac{1}{12}+\left(\frac{1}{2^4}+\frac{1}{3\times 2^5}+\frac{1}{3^2\times 2^6}+\&c.\right)\right\}=D\times\left(\frac{1}{12}+\frac{3}{40}\right)=D\times 0\cdot 16.$$

So that each grandparent contributes on the average $0\cdot 16$ (more exactly $0\cdot 1583$) of his peculiarity to the heritage of S, and the four grandparents contribute between them $0\cdot 64$, leaving 36 indeterminate, which when rateably assigned gives $0\cdot 25$ to light and $0\cdot 11$ to dark. A hazel-eyed grandparent contributes, according to the ratio described in the last paragraph, $0\cdot 10$ to light and $0\cdot 06$ to dark. All this is clearly expressed and employed in the columns II. of Tables 17 and 18.

III. Eye-colours given of the two Parents and four Grand-Parents—

Suppose F to possess D, then F taken alone, and not in connection with what his possession of D might imply concerning the contributions of the previous ancestry, will contribute an average of $0\cdot 25$ to the heritage of S. Suppose G_1 also to possess D, then his contribution together with what his possession of D may imply concerning the previous ancestry, was calculated in the last paragraph as $D\times\frac{3}{40}=D\times 0\cdot 075$. For the convenience of using round numbers I take this as $D\times 0\cdot 08$. So the two parents contribute between

them 0·50 of the peculiarity of S, the four grandparents together with what they imply of the previous
ancestry contribute 0·32, being an aggregate of 0·82,
leaving a residue of 0·18 to be rateably assigned as
0·12 to light, and 0·6 to dark. A hazel-eyed Parent
is here reckoned as contributing 0·16 to light and 0·9 to
dark; a hazel-eyed Grand-Parent as contributing 0·5
to light and 0·3 to dark. All this is tabulated in
Table 17, and its working explained by an example in
the columns headed III. of Table 18.

Results.—A mere glance at Tables 19 and 20 will
show how surprisingly accurate the predictions are, and
therefore how true the basis of the calculations must be.
Their *average* correctness is shown best by the totals
in Table 19, which give an aggregate of calculated
numbers of light-eyed children under Groups I., II., and
III. as 623, 601, and 614 respectively, when the observed
numbers were 629 ; that is to say, they are correct in
the ratios of 99, 96, and 98 to 100.

Their trustworthiness when applied to *individual*
families is shown as strongly in Table 20 whose results
are conveniently summarised in Table 21. I have there
classified the amounts of error in the several calculations :
thus if the estimate in any one family was 3 light-
eyed children, and the observed number was 4, I should
count the error as 1·0. I have worked to one place of
decimals in this table, in order to bring out the different
shades of trustworthiness in the three sets of calculations, which thus become very apparent. It will be

seen that the calculations in Class III. are by far the most precise. In more than one-half of those calculations the error does not exceed 0·5, whereas in more than three-quarters of those in I. and II. the error is at least of that amount. Only one-quarter of Class III., but somewhere about the half of Classes I. and II., are more than 1·1 in error. In comparing I. with II., we find I. to be slightly but I think distinctly the superior estimate. The relative accuracy of III. as compared with I. and II., is what we should have expected, supposing the basis of the calculations to be true, because the additional knowledge utilised in III., over what is turned to account in I. and II., must be an advantage.

My returns are insufficiently numerous and too subject to uncertainty of observation, to make it worth while to submit them to a more rigorous analysis, but the broad conclusion to which the present results irresistibly lead, is that the same peculiar hereditary relation that was shown to subsist between a man and each of his ancestors in respect to the quality of Stature, also subsists in respect to that of Eye-colour.

CHAPTER IX

THE ARTISTIC FACULTY.

Data.—Sexual Distribution.—Marriage Selection.—Regression.—Effect of
Bias in Marriage.

Data.—It is many years since I described the family
history of the great Painters and Musicians in *Hereditary Genius.* The inheritance of much less exceptional gifts of Artistic Faculty will be discussed in this
chapter, and from an entirely different class of data.
They are the answers in my R.F.F collection, to the
question of "Favourite pursuits and interests ? Artistic
aptitudes ? "

The list of persons who were signalised as being
especially fond of music and drawing, no doubt
includes many who are artistic in a very moderate
degree. Still they form a fairly well defined class,
and one that is easy to discuss because their family
history is complete. In this respect, they are much
more suitable subjects for statistical inquiry than the
great Painters and Musicians, whose biographers usually
say little or nothing of their non-artistic relatives.

The object of the present chapter is not to give a reply to the simple question, whether or no the Artistic faculty tends to be inherited. A man must be very crotchety or very ignorant, who nowadays seriously doubts the inheritance either of this or of any other faculty. The question is whether or no its inheritance follows a similar law to that which has been shown to govern Stature and Eye-colour, and which has been worked out with some completeness in the foregoing chapters. Before answering this question, it will be convenient to compare the distribution of the Artistic faculty in the two sexes, and to learn the influence it may exercise on marriage selection.

I began by dividing my data into four classes of aptitudes; the first was for music alone; the second for drawing alone; the third for both music and drawing; and the fourth includes all those about whose artistic capacities a discreet silence was observed. After prefatory trials, I found it so difficult to separate aptitude for music from aptitude for drawing, that I determined to throw the three first classes into the single group of Artistic. This and the group of the Non-Artistic are the only two divisions now to be considered.

A difficulty presented itself at the outset in respect to the families that included boys, girls, and young children, whose artistic tastes and capacities can seldom be fairly judged, while they are liable to be appraised too favourably by the compiler of the Family records, especially if he or she was one of their parents. As the practice of picking and choosing is very hazardous in statistical inquiries, however fair our intentions may be, and as it in justice always excites suspicion, I decided, though with much regret at their loss, to omit the whole of those who were not adult.

Sexual Distribution.—Men and women, as classes, may differ little in their natural artistic capacity, but such difference as there is in adult life is somewhat in favour of the women. Table 9B contains 894 cases, 447 of men and 447 of women, divided into three groups according to the rank they hold in the pedigrees. These groups agree fairly well among themselves, and therefore their aggregate results may be freely accepted as trustworthy. They show that 28 per cent. of the males are Artistic and 72 are Not Artistic, and that there are 33 per cent. Artistic females to 67 who are Not Artistic. Part of this female superiority is doubtless to be ascribed to the large share that music and drawing occupy in the education of women, and to the greater leisure that most girls have, or take, for amusing themselves. If the artistic gifts of men and women are naturally the same, as the experience of schools where music and drawing are taught, apparently shows it to be, the small difference observed in favour of women in adult life would be a measure of the smallness of the effect of education compared to that of natural talent. Disregarding the distinction of sex, the figures in Table 9B show that the number of Artistic to Non-Artistic persons in the general population is in the proportion of 30½ to 69½. The data used in Table 22 refer to a considerably larger number of persons, and do not include more than two-thirds of those employed in Table 9B, and they make the proportion to be 31 to 69. So we shall be quite correct enough if we reckon that out of ten persons in the families of my R.F.F. correspondents, three on the average are artistic and seven are not.

Marriage Selection.—Table 9B enables us to ascertain whether there is any tendency, or any disinclination among the Artistic and the Non-Artistic, to marry within their respective castes. It shows the observed frequency of their marriages in each of the three possible combinations; namely, both husband and wife artistic; one artistic and one not; and both not artistic. The Table also gives the calculated frequency of the three classes, supposing the pairings to be regulated by the laws of chance. There is I think trustworthy evidence of the existence of some slight disinclination to marry within the same caste, for signs of it appear in each of the three sets of families with which the Table deals. The total result is that there are only 36 per cent. of such marriages observed, whereas if there had been no disinclination but perfect indifference, the number would have been raised to 42. The difference is small and the figures are few, but for the above reasons it is not likely to be fallacious. I believe the facts to be, that highly artistic people keep pretty much to themselves, but that the very much larger body of moderately artistic people do not. A man of highly artistic temperament must look on those who are deficient in it, as barbarians; he would continually crave for a sympathy and response that such persons are incapable of giving. On the other hand, every quiet unmusical man must shrink a little from the idea of wedding himself to a grand piano in constant action, with its vocal and peculiar social accompaniments; but he might anticipate great pleasure in having a wife of a moderately artistic temperament, who would give colour and variety to his prosaic life. On the other hand, a sensitive and imaginative wife would be conscious of needing the aid of a husband who had enough plain common-sense to restrain her too enthusiastic and frequently foolish projects. If wife is read for husband, and husband for wife, the same argument still holds true.

Regression.—Having disposed of these preliminaries, we will now examine into the conditions of the inheritance of the Artistic Faculty. The data that bear upon it are summarised in Table 22, where I have not cared to separate the sexes, as my data are not numerous enough to allow of more subdivision than can be helped. Also, because from such calculations as I have made, the hereditary influences of the two sexes in respect to art appear to be pretty equal : as they are in respect to nearly every other characteristic, exclusive of diseases, that I have examined.

It is perfectly conceivable that the Artistic Faculty

in any person might be somehow measured, and its amount determined, just as we may measure Strength, the power of Discrimination of Tints, or the tenacity of Memory. Let us then suppose the measurement of the Artistic Faculty to be feasible and to have been often performed, and that the measures of a large number of persons were thrown into a Scheme.

It is reasonable to expect that the Scheme of the Artistic Faculty would be approximately Normal in its proportions, like those of the various Qualities and Faculties whose measures were given in Tables 2 and 3.

It is also reasonable to expect that the same law of inheritance might hold good in the Artistic Faculty that was found to hold good both in Stature and in Eye colour; in other words, that the value of Filial Regression would in this case also be $\frac{2}{3}$.

We have now to discover whether these assumptions are true without any help from direct measurement. The problem to be solved is a pretty one, and will illustrate the method by which many problems of a similar class have to be worked.

Let the graduations of the scale by which the Artistic Faculty is supposed to be measured, be such that the unit of the scale shall be equal to the Q of the Art-Scheme of the general population. Call the unknown M of the Art-Scheme of the population, P. Then, as explained in page 52, the measure of any individual will be of the form P + (± D), where D is the deviation from P. The first fact we have to deal with is, that only 30 per cent. of the population are Artistic.   Therefore no person whose Grade in the Art-Scheme does not exceed 70° can be reckoned as Artistic.   Referring to Table 8 we see that the value of D for the Grade of 70° is + 0·78 ; consequently the art-measure of an Artistic person, when reckoned in units of the accepted scale, must exceed $P + 0·78$.

The average art-measure of all persons whose Grade is higher than 70°, may be obtained with sufficient approximation by taking the average of all the values given in Table 8, for every Grade, or more simply for every odd Grade from 71° to 99° inclusive.   It will be found to be 1·71.   Therefore an artistic person has, on the average, an art-measure of $P + 1·71$.   We will consider persons of this measure to be representatives of the whole of the artistic portion of the Population.   It is not strictly correct to do so, but for approximative purposes this rough and ready method will suffice, instead of the tedious process of making a separate calculation for each Grade.

The M of the Co-Fraternity born of a group of Mid-Parents whose measure is $P + 1·71$ will be $P + (\frac{2}{3} \times 1·71)$ or $(P + 1·4)$.   We will call this value C.   The Q of this or any other Co-Fraternity may be expected to bear approximately the same ratio to the Q of the general population, that it did in the case of Stature, namely, that of 1·5 to 1·7.   Therefore the Q of the Co-Fraternity who are born of Mid-Parents whose Art-measure is C, will be 0·88.

The artistic members of this Co-Fraternity will be those whose measures exceed $\{P + 0·78\}$.   We may write this

value in the form of $\{(P + 1\cdot4) - 0\cdot36\}$, or $\{C - 0\cdot36\}$.
Table 8 shows that the Deviation of $-0\cdot36$ is found
at the Grade of 40°. Consequently 40 per cent. of
this Co-Fraternity will be Non-Artistic and 60 per cent.
will be Artistic. Observation Table 23 shows the
numbers to be 36 and 64, which is a very happy
agreement.

Next as regards the Non-Artistic Parents. The Non-
Artistic portion of the Population occupy the 70 first
Grades in the Art-Scheme, and may be divided into
two groups; one consisting of 40 Grades, and standing
between the Grades of 70° and 30°, or between the
Grade of 50° and 20 Grades on either side of it, the
average Art-measure of whom is P; the other group
standing below 30°, whose average measure may be taken
to be $P - 1\cdot71$, for the same reason that the group
above 70° was taken as $P + 1\cdot71$. Consequently the
average measure of the entire Non-Artistic class is

$$\tfrac{1}{70} \{(40 \times P) + 30 (P - 1\cdot71)\}$$
$$= P - \tfrac{30}{70} \times 1\cdot71 = P - 0\cdot73.$$

Supposing Mid-Parents of this measure, to represent the
entire Non-Artistic group, their offspring will be a Co-
Fraternity having for their M the value of $P - \{\tfrac{2}{3} \times 0\cdot73\}$
or $P - 0\cdot49$, which we will call C', and for their Q the
value of $0\cdot88$ as before.

Such among them as exceed $\{P - 0\cdot78\}$, which we
may write in the form of $\{(P - 0\cdot49) + (1\cdot27)\}$, or
$\{C' + 1\cdot27\}$, are Artistic, and they are those who,
according to Table 8, rank higher than the Grade 83°.
In other words, 83 per cent. of the children of Non-

Artistic parents will be Non-Artistic, and the remainder of 17 per cent. will be Artistic. Observation
gives the values of 79 and 21, which is a very fair
coincidence.

When one parent is Artistic and the other Not, their
joint hereditary influence would be the average of the
above two cases; that is to say, $\frac{1}{2}$ (40 + 83), or $61\frac{1}{2}$
per cent. of their children would be Non-Artistic, and
$\frac{1}{2}$ (60 + 17), or $38\frac{1}{2}$, would be Artistic. The observed
numbers are 61 and 39, which agree excellently well.

We may therefore conclude that the same law of
Regression, and all that depends upon it, which governs
the inheritance both of Stature and Eye-colour, applies
equally to the Artistic Faculty.

Effect of Bias in Marriage.—The slight apparent
disinclination of the Artistic and the Non-Artistic to
marry in their own caste, is hardly worth regarding,
but it is right to clearly understand the extreme effect
that might be occasioned by Bias in Marriage. Suppose
the attraction of like to like to become paramount, so
that each individual in a Scheme married his or her
nearest available neighbour, then the Scheme of Mid-
Parents would be practically identical with the Scheme
drawn from the individual members of the population.
In the case of Stature their Q would be 1·7 inch, instead
of 1·7 divided by $\sqrt{2}$. The regression and subsequent
dispersion remaining unchanged, the Q of the offspring
would consequently be increased.

On the other hand, suppose the attraction of contrast

to become suddenly paramount, so that Grade 99° paired in an instant with Grade 1°; next 98° with 2°; and so on in order, until the languid desires of 49° and 51° were satisfied last of all. Then every one of the Mid-Parents would be of precisely the same stature P. Consequently their Q would be zero; and that of the system of the Mid-Co-Fraternities would be zero also; hence the Q of the next generation would contract to the Q of a Co-Fraternity, that is to 1·5 inch.

Whatever might be the character or strength of the bias in marriage selection, so long as it remains constant the Q of the population would tend to become constant also, and the statistical resemblance between successive generations of the future Population would be ensured. The stability of the balance between the opposed tendencies of Regression and of Co-Fraternal expansion is due to the Regression increasing with the Deviation. Its effect is like that of a spring acting against a weight; the spring stretches until its gradually increasing resilient force balances the steady pull of the weight, then the two forces of spring and weight are in stable equilibrium. For, if the weight be lifted by the hand, it will obviously fall down again as soon as the hand is withdrawn; or again, if it be depressed by the hand, the resilience of the spring will become increased, and the weight will rise up again when it is left free to do so.

CHAPTER X.

DISEASE.

Preliminary Problem.—Data.—Trustworthiness of R.F.F. Data.—Mixture of Inheritances.—CONSUMPTION: General Remarks; Parent to Child; Distribution of Fraternities; Severely Tainted Fraternities; Consumptivity.—Data for Hereditary Diseases.

THE vital statistics of a population are those of a vast army marching rank behind rank, across the treacherous table-land of life. Some of its members drop out of sight at every step, and a new rank is ever rising up to take the place vacated by the rank that preceded it, and which has already moved on. The population retains its peculiarities although the elements of which it is composed are never stationary, neither are the same individuals present at any two successive epochs. In these respects, a population may be compared to a cloud that seems to repose in calm upon a mountain plateau, while a gale of wind is blowing over it. The outline of the cloud remains unchanged, although its elements are in violent movement and in a condition of perpetual destruction and renewal. The

well understood cause of such clouds is the deflection
of a wind laden with invisible vapour, by means of
the sloping flanks of the mountain, up to a level at
which the atmosphere is much colder and rarer than
below. Part of the invisible vapour with which the
wind was charged, becomes thereby condensed into the
minute particles of water of which clouds are formed.
After a while the process is reversed. The particles
of cloud having been carried by the wind across the
plateau, are swept down the other side of it again to a
lower level, and during their descent they return into
invisible vapour. Both in the cloud and in the
population, there is on the one hand a continual supply
and inrush of new individuals from the unseen ; they
remain a while as visible objects, and then disappear.
The cloud and the population are composed of elements
that resemble each other in the brevity of their existence, while the general features of the cloud and of the
population are alike in that they abide.

Preliminary Problem.—The proportion of the
population that dies at each age, is well known, and the
diseases of which they die are also well known, but the
statistics of hereditary disease are as yet for the most
part contradictory and untrustworthy.

It is most desirable as a preliminary to more minute
inquiries, that the causes of death of a large number of
persons should be traced during two successive generations in somewhat the same broad way that Stature
and several other peculiarities were traced in the pre-

ceding chapters. There are a certain number of recognized groups of disease, which we may call A; B, C, &c.,
and the proportion of persons who die of these diseases
in each of the two generations is the same. The preliminary question to be determined is whether and to
what extent those who die of A in the second generation, are more or less often descended from those who
died of A in the first generation, than would have been
the case if disease were neither hereditarily transmitted
nor clung to the same families for any other reason.
Similarly as regards B, C, D, and the rest.

This inquiry would be more difficult than those
hitherto attempted, because longevity and fertility are
both affected by the state of health, and the circumstances of home life and occupation have a great effect
in causing and in checking disease. Also because the
father and mother are found in some notable cases to
contribute disease in very different degrees to their
male and female descendants.

I had hoped even to the last moment, that my
collection of Family Records would have contributed
in some small degree towards answering this question,
but after many attempts I find them too fragmentary
for the purpose. It was a necessary condition of success
to have the completed life-histories of many Fraternities
who were born some seventy or more years ago, that
is, during the earlier part of this century, as well as
those of their parents and all their uncles and aunts.
My Records contain excellent material of a later date,
that will be valuable in future years; but they must

bide their time ; they are insufficient for the period in question. By attempting to work with incompleted life histories the risk of serious error is incurred.

Data.—The Schedule in Appendix G, which is illustrated in more detail by Tables A and B that follow it, shows the amount of information that I had hoped to obtain from those who were in a position to furnish complete returns. It relates to the "Subject" of the pedigree and to each of his 14 direct ancestors, up to the great-grandparents inclusive, making in all 15 persons. Also, to the Fraternities of which each of these 15 persons was a member. Reckoning the total average number of persons in each fraternity at 5, which is under the mark for my R.F.F collection, questions were thus asked concerning an average of 75 different persons in each family. The total number of the Records that I am able to use, is about 160; so the aggregate of the returns of disease ought to have been about twelve thousand, and should have included the causes of death of perhaps 6,000 of them. As a matter of fact, I have only about one-third of the latter number.

Trustworthiness of R.F.F. data.—The first object was to ascertain the trustworthiness of the medical information sent to me. There is usually much disinclination among families to allude to the serious diseases that they fear to inherit, and it was necessary to learn whether this tendency towards suppression notably vitiated the returns. The test applied was both simple and just.

If consumption, cancer, drink and suicide, appear among the recorded cases of death less frequently than they do in ordinary tables of mortality, then a bias towards suppression could be proved and measured, and would have to be reckoned with ; otherwise the returns might be accepted as being on the whole honest and outspoken. I find the latter to be the case. Sixteen per cent. of the causes of death (or 1 in 6½) are ascribed to consumption, 5 per cent. to cancer, and nearly 2 per cent. to drink and to suicide respectively. Insanity was not specially asked about, as I did not think it wise to put too many disagreeable questions, however it is often mentioned. I dare say that it, or at least eccentricity, is not unfrequently passed over. Careful accuracy in framing the replies appears to have been the rule rather than the exception. In the preface to the blank forms of the *Records of Family Faculties* and elsewhere, I had explained my objects so fully and they were so reasonable in themselves, that my correspondents have evidently entered with interest into what was asked for, and shown themselves willing to trust me freely with their family histories. They seem generally to have given all that was known to them, after making much search and many inquiries, and after due references to registers of deaths. The insufficiency of their returns proceeds I feel sure, much less from a desire to suppress unpleasant truths than from pure ignorance, and the latter is in no small part due to the scientific ineptitude of the mass of the members of the medical profession two and more generations ago, when even the stetho-

scope was unknown. They were then incompetent to
name diseases correctly.

Mixture of Inheritances.—The first thing that struck
me after methodically classifying the diseases of each
family, in the form shown in the Schedule, was their
great intermixture.   The Tables A and B in Appendix G
are offered as ordinary specimens of what is everywhere
to be found.   They are actual cases, except that I have
given fancy names and initials, and for further conceal-
ment, have partially transposed the sexes.   Imagine an
intermarriage between any two in the lower division of
these tables, and then consider the variety of inheritable
disease to which their children would be liable !   The
problem is rendered yet more complicated by the
metamorphoses of disease.   The disease A in the parent
does not necessarily appear, even when inherited, as A in
the children.   We know very little indeed about the
effect of a mixture of inheritable diseases, how far they
are mutually exclusive and how far they blend ; or how
far when they blend, they change into a third form.
Owing to the habit of free inter-marriage no person can
be exempt from the inheritance of a vast variety of
diseases or of special tendencies to them.   Deaths by
mere old age and the accompanying failure of vital
powers without any well defined malady, are very
common in my collection, but I do not find as a rule,
that the children of persons who die of old age have any
marked immunity from specific diseases.

There is a curious double appearance in the Records.

the one of an obvious hereditary tendency to disease
and the other of the reverse. There are far too many
striking instances of coincidence between the diseases of
the parents and of the children to admit of reasonable
doubt of their being often inherited. On the other hand,
when I hide with my hand the lower part of a page such
as those in Tables A and B, and endeavour to make
a forecast of what I shall find under my hand after
studying the upper portion, I am sometimes greatly mistaken. Very unpromising marriages have occasionally
led to good results, especially where the parental disease
is one that usually breaks out late in life, as in the case
of cancer. The children may then enjoy a fair length
of days and die in the end of some other disease;
although if that disease had been staved off it is quite
possible that the cancer would ultimately have appeared.
I have two remarkable instances of this. In one of
them, three grandparents out of four died of cancer. In
each of the fraternities of which the father and mother
were members, one and one person only, died of it.
As to the children, although four of them have lived to
past seventy years, not one has shown any sign of
cancer. The other case differs in details, but is equally
remarkable. However diseased the parents may be, it
is of course possible that the children may inherit the
healthier constitutions of their remoter ancestry. Pro-
mising looking marriages are occasionally found to lead
to a sickly progeny, but my materials are too scanty to
permit of a thorough investigation of these cases.

The general conclusion thus far is, that owing to

the hereditary tendencies in each person to disease being usually very various, it is by no means always that useful forecasts can be made concerning the health of the future issue of any couple.

CONSUMPTION.

General Remarks.—The frequency of consumption in England being so great that one in at least every six or seven persons dies of it, and the fact that it usually appears early in life, and is therefore the less likely to be forestalled by any other disease, render it an appropriate subject for statistics. The fact that it may be acquired, although there has been no decided hereditary tendency towards it, introduces no serious difficulty, being more or less balanced by the opposite fact that it may be withstood by sanitary precautions although a strong tendency exists. Neither does it seem worth while to be hypercritical and to dwell overmuch on the different opinions held by experts as to what constitutes consumption. The ordinary symptoms are patent enough, and are generally recognized; so we may be content at first with lax definitions. At the same time, no one can be more strongly impressed than myself with the view that in proportion as we desire to improve our statistical work, so we must be increasingly careful to divide our material into truly homogeneous groups, in order that all the cases contained in the same group shall be alike in every important particular, differing only in petty details. This is far more important than adding to the number of cases. My material admits of no such delicacy of division; nevertheless it leads to some results worth mentioning.

In sorting my cases, I included under the head of Consumption all the causes of death described by one or the other following epithets, attention being also paid to the context, and to the phraseology used elsewhere by the same writer:—Consumption; Phthisis; Tubercular disease; Tuberculosis; Decline; Pulmonary, or lung disease; Lost lung; Abscess on lung; Hæmorrhage of lungs (fatal); Lungs affected (here especially the context was considered). All of these were reckoned as actual Consumption.

In addition to these there were numerous phrases of doubtful import that excited more or less reasonable suspicion. It may be that the disease had not sufficiently declared itself to justify more definite language, or else that the phrase employed was a euphemism to veil a harsh truth. Paying still more attention to the context than before, I classed these doubtful cases under three heads:—(1) Highly suspicious; (2) Suspicious; (3) Somewhat suspicious. They were so rated that four cases of the first should be reckoned equivalent to three cases of actual consumption, four cases of the second to two cases, and four of the third to one case.

The following is a list of some of the phrases so dealt with. The occasional appearance of the same phrase under different headings is due to differences in the context :—

1. *Highly suspicious*:—Consumptive tendency, Con-

sumption feared, and died of bad chill. Chest colds
with pleurisy and congestion of lungs. Died of an
attack on the chest. Always delicate. Delicate lungs.
Hæmorrhage of lungs. Loss of part of lung. Severe
pulmonary attacks and chest affections.

2. *Suspicious* :—Chest complaints. Delicate chest.
Colds, cough and bronchitis. Delicate, and died of
asthma. Scrofulous tendency.

3. *Somewhat suspicious* :—Asthma when young. Pulmonary congestion. Not strong; anæmic. Delicate.
Colds, coughs. Debility; general weakness. [The context was especially considered in this group.]

Parent to Child.—I have only four cases in which both
parents were consumptive; these will be omitted in the following remarks ; but whether included or not, the results
would be unaltered, for they run parallel to the rest.

There are 66 marriages in which one parent was
consumptive ; they produced between them 413 children, of whom 70 were actually consumptive, and others
who were suspiciously so in various degrees. When
reckoned according to the above method of computation,
these amounted to 37 cases in addition, forming a total
of 107. In other words, 26 per cent. of the children
were consumptive. Where neither parent was consumptive, the proportion in a small batch of well marked
cases that I tried, was as high as 18 or 19 per cent., but
this is clearly too much, as that of the general population
is only 16 per cent. Again, by taking each fraternity
separately and dividing the quantity of consumption in
it by the number of its members, I obtained the average

consumptive taint of each fraternity. For instance, if
in a fraternity of 10 members there was one actually
consumptive member and four "somewhat suspiciously"
so, it would count as a fraternity of ten members, of
whom two were actually consumptive, and the average
taint of the fraternity would be reckoned at one-fifth
part of the whole or as 20 per cent.

Treating each fraternity separately in this way, and
then averaging the whole of them, the mean taint of
the children of one consumptive parent was made out
to be 28 per cent.

Distribution of Fraternities.—Next I arranged the
fraternities in such way as would show whether, if we
reckoned each fraternity as a unit, their respective
amounts of consumptive taint were distributed "normally" or not.   The results are contained in line A of
the following table :—

PERCENTAGE OF CASES HAVING VARIOUS PERCENTAGES OF TAINT.

	Percentages of Taint.					
	0 and under 9	10 and under 19	20 and under 29	30 and under 39	40 and above	Total.
A. 66 cases, one parent consumptive.	27	20	9	15	29	100
B. 84 cases, one brother consumptive.	49	14	10	13	14	100

They struck me as so remarkable, in the way shortly
to be explained, that I proceeded to verify them by as
different a set of data as my Records could afford.   I
took every fraternity in which at least one member
was consumptive, and treated them in a way that would
answer the following question.   " One member of a
fraternity, whose number is unknown, is consumptive ;
what is the chance that a named but otherwise unknown brother of that man will be consumptive also ? "
The fraternity that was taken above as an example,
would be now reckoned as one of nine members, of
whom one was actually consumptive.   There were 84
fraternities available for the present purpose, and the
results are given in the line B of the table.   The data
in A and B somewhat overlap, but for the most part
they differ.

They concur in telling the same tale, namely, that it
is totally impossible to torture the figures so as to make
them yield the single-humped " Curve of Frequency "
(Fig. 3 p. 38).   They make a distinctly double-humped
curve, whose outline is no more like the normal curve
than the back of a Bactrian camel is to that of an
Arabian camel.   Consumptive taints reckoned in this
way are certainly not " normally " distributed.   They
depend mainly on one or other of two groups of causes,
one of which tends to cause complete immunity and
the other to cause severe disease, and these two groups
do not blend freely together.   Consumption tends to
be transmitted strongly or not at all, and in this respect
it resembles the baleful influence ascribed to cousin

marriages, which appears to be very small when
statistically discussed, but of whose occasional severity
most persons have observed examples.

I interpret these results as showing that consumption
is largely acquired, and that the hereditary influence of
an acquired attack is small when there is no accompanying "malformation." This last phrase is intended
to cover not only a narrow chest and the like, but whatever other abnormal features may supply the physical
basis upon which consumptive tendencies depend, and
which I presume to be as hereditary as any other
malformations.

Severely-tainted Fraternities.—Pursuing the matter
further, I selected those fraternities in which consumption was especially frequent, and in which the causes of
the deaths both of the Father and of the Mother were
given. They were 14 in number, and contained between them a total of 102 children, of whom rather
more than half died before the age of 40. Though
records of infant deaths were asked for, I doubt if
they have been fully supplied. As 102 differs little
from 100, the following figures will serve as percentages : 42 died of actual consumption and 11 others
of lung disease variously described. Only one case
was described as death from heart disease, but weakness
of the heart during life was spoken of in a few cases.
The remaining causes of death were mostly undescribed,
and those that were named present no peculiarity worth
notice. I then took out the causes of death of the

Fathers and Mothers and their ages at death, and severally classified them as in the Table below. It must be understood that there is nothing in the Table to show how the persons were paired. The Fathers are treated as a group by themselves, and the Mothers as a separate group, also by themselves.

CAUSES OF DEATH OF THE PARENTS OF THOSE FRATERNITIES IN WHICH CONSUMPTION GREATLY PREVAILED.

Father.	Age at death.	Mother.	Age at death.	Order of ages at death. F.	Order of ages at death. M.
Asthma	70	Consumption	40	51	40
Bronchitis	89	Consumption	43	52	42
Inf. kidneys and bronchitis .	73	Consumption	47	59	43
Abscess of liver through lung(alive)		Consumption	55	62	44
Heart	68	Consumption	66	68	47
Heart	74	Consumption	66	70	50
Apoplexy.	62	Water on chest	60	73	58
Apoplexy.	75	Weak chest. . . . (alive)		74	60
Apoplexy.	78	(1 br. and 2 ss. d. of cons.)		74	65
Decay	74	Hæmorrhage of lungs .	44	75	66
Cancer	52	Ossification of heart . .	50	76	73
Senile gangrene	76	Nose bleeding.	83	78	74
(2 bros. d. of cancer).		Cancer	42	89	83
Mortification of toe	59	Atrophy	73		
Accident	51	Age	74		
(3 bros. and 2 ss. d. of cons.)					

Very little account is given of the fraternities to which the fathers and mothers belong, and nothing of interest beyond what is included in the above.

The contrast is here most striking between the tendencies of the Father and Mother to transmit a serious consumptive taint to their children. The cases were selected without the slightest bias in favour of showing this result; in fact, such is the incapacity to see statistical facts clearly until they are pointed out, that I had no idea of the extraordinary tendency on

N

the part of the mother to transmit consumption, as
shown in this Table, until I had selected the cases and
nearly finished sorting them.   Out of the fourteen
families, the mother was described as actually dying
of consumption in six cases, of lung complaints in two
others, and of having highly consumptive tendencies
in another, making a total of nine cases out of the
fourteen.   On the other hand the Fathers show hardly
any consumptive taints.   One was described as of a
very consumptive fraternity, though he himself died of
an accident ; and another who was still alive had suffered
from an abscess of the liver that broke through the
lungs.   Beyond these there is nothing to indicate
consumption on the Fathers' side.

Another way of looking at the matter is to compare
the ages at death of the Mothers and of the Fathers
respectively, as has been done at the side of the Table,
when we see a notable difference between them, the
Mid-age of the Mothers being 58, as against 73 of
the Fathers.

The only other group of diseases in my collection,
that affords a fair number of instances in which fraternities are greatly affected, are those of the Heart.
The instances are only nine in number, but I give an
analysis of them, not for any value of their own, but
in order to bring the peculiarities of the consumptive
fraternities more strongly into relief by means of comparison.   In one of these there was no actual death
from heart disease, though three had weak hearts and
two others had rheumatic gout and fever.   These nine

families contained between them sixty-nine children, being at the rate of 7·7 to a family. The number of deaths from heart disease was 24 ; from ruptured blood vessels, 2 ; from consumption and lung disease, 8 ; from dropsy in various forms, 3 ; from apoplexy, paralysis, and epilepsy, 5 ; from suicide, 2 ; from

CAUSES OF DEATH OF THE PARENTS OF THOSE FRATERNITIES IN WHICH HEART DISEASE PREVAILED.

Causes of death.	Ages at death.		Order of ages at death.	
	Father.	Mother.	F.	M.
Heart	59, 70	61, 63, 74	53	61
Apoplexy and paralysis .	74, 78	62, 70, 72	55	62
Consumption	53	....	59	63
Asthma	70	...	70	70
Gout	65	...	70	72
Senile Gangrene	...	81	74	74
Tumour in liver	..	77	75	77
Cancer	75	...	78	81
Living	old.	...	old.	85
Unknown		85		
		2 bros. and 1 sis. d. of heart disease and 1 of paralysis æt. 40.		

cancer, 1. There is no obvious difference between the diseases of their Fathers and Mothers as shown in the Table, other than the smallness of the number of cases would account for. Their mid-ages at death were closely the same, 70 and 72, and the ages in the two groups run alike.

I must leave it to medical men to verify the amount of truth that may be contained in what I have deduced from these results concerning the distinctly superior

power of the mother over that of the father to produce
a highly consumptive family.   Any physician in large
practice among consumptive cases could test the question easily by reference to his note-books.   A "highly
consumptive" fraternity may conveniently be defined
as one in which at least half of its members have
actually died of consumption, or else are so stricken
that their ultimate deaths from that disease may be
reckoned upon.   Also to avoid statistical accidents, the
fraternities selected for the inquiry should be large,
consisting say of six children and upwards.   Of course
the numerical proportions given by the above 14 fraternities are very rude indications indeed of the results to
which a thorough inquiry might be expected to lead.

Accepting the general truth of the observation
that consumptive mothers produce highly consumptive
families much more commonly than consumptive fathers,
it is easy to offer what seems to be an adequate explanation.   Consumption is partly acquired by some
form of contagion or infection, and is partly an hereditary malformation.   So far as it is due to the latter
in the wide sense already given to the word "malformation," it may perhaps be transmissible equally by
either parent.   But so far as it is contagious or
infectious, we must recollect that the child is pecu-
liarly exposed during all the time of its existence
before birth, to contagion from its mother.   During
infancy, it lies perhaps for hours daily in its mother's
arms, and afterwards lives much by her side, closely
caressed, and breathing the tainted air of her sheltered

rooms. The explanation of the fact that we have been discussing appears therefore to be summed up in the single word—Infection.

Consumptivity. — Before abandoning the topic of hereditary consumption, it may be well to discuss it from the same point of view that was taken when discussing the artistic temperament. Consumption being so common in this country that fully one person out of every six or seven die of it, and all forms of hereditary disease being intermixed through marriage, it follows that the whole population must be more or less tainted with consumption. That a condition which we may call "consumptivity," for want of a better word, may exist without showing any outward sign, is proved by the fact that as sanitary conditions worsen by ever so little, more persons are affected by the disease. It seems a fair view to take, that when the amount of consumptivity reaches a certain level, the symptoms of consumption declare themselves; that when it approaches but falls a little short of that level, there are threatening symptoms; that when it falls far below the level, there is a fallacious appearance of perfect freedom from consumptivity. We may reasonably proceed on the hypothesis that consumptivity might somehow be measured, and that if its measurement was made in each of any large group of persons, the measures would be distributed "normally."

So far we are on fairly safe ground, but now uncertainties begin upon which my data fail to throw sufficient light. Longevity, marriage, and fertility must all be affected by the amount of consumptivity, whereas in the case of the faculties hitherto discussed they are not affected to any sensible extent. It however happens that these influences tend to neutralize one another. It is true that consumptive persons die early, and many of them before a marriageable age. On the other hand, they certainly marry earlier as a rule than others, one cause of which lies in their frequent great attractiveness; and again, when they marry, they produce children more quickly than others. Consequently those who die even long before middle age, often contrive to leave large families. The greater rapidity with which the generations follow each other, is also a consideration of some importance. There is therefore a fair doubt whether a group of young persons destined to die of consumption, contribute considerably less to the future population than an equally large group who are destined to die of other diseases. I will at all events assume that consumptivity does not affect the numbers of the adult children, simply as a working hypothesis, and will afterwards compare its results with observed facts.

I should add that the question whether the sexes transmit consumption equally, lies outside the present work, at least for practical purposes; for whether they transmit it equally or not would not affect the results materially. Our list of data is therefore limited to these :—that 16 per cent. of the population die of consumption, that consumptivity is normally distri-

buted, and that the law of hereditary regression from
a deviation of three units on the part of either
parent to an average of one unit in the child, may
be supposed to apply here, just as it did to Stature
and to the other subjects of the preceding chapters.

Let the scale by which consumptivity is measured be
such that the Q of the general population = 1. Let
its M = N, when measured on the same scale ; the
value of N is and will remain unknown. Let N + C
be the number of units of consumptivity that just
amount to actual consumption. Our data tell us that
16 per cent. of the population have an amount of consumptivity that exceeds N + C. On referring to
Table 8, we find the value of C that corresponds to the
Grade of (100°—16°), or of 84°, to be 1·47. Therefore whenever the consumptivity of a person exceeds
N + 1·47, he has actual consumption.

Adding together the tabular values in Table 8 at all
the odd grades above 84°, we shall find their average
value to be 2·23. We may therefore assume (see p. 160)
that a group of persons each of whom has a consumptivity of N + 2·23 will approximately represent all the
grades above 84°. The Co-Fraternity descended from
such a group will have an M whose value according
to the law of Regression ought to be $[N + \frac{1}{3} (2·23)]$
or $[N + 0·74$ units.$]$

Those members of the Co-Fraternity are consumptive
whose consumptivity exceeds N + 1·47 ; these are the
same as those whose deviation from $[N + 0·74]$ which
is the M of the Co-Fraternity, exceeds + 0·73 unit.

Let the Q of the Co-Fraternity be called n. The Grade at which this amount of deviation occurs should be found in Table 8 opposite to the value of 0·73 divided by n.

Next as regards the value to be assigned to n, we may be assured that the Q of a Co-Fraternity cannot exceed that of the general population. Therefore n cannot exceed 1. In the case of Stature the relation between the Q of the Co-Fraternity and that of the Population was found to be as 15 to 17. If the same proportion held good here, its value would be 0·9. This is I think too high an estimate for the following reasons. The variability of the Co-Fraternity depends on two groups of causes. First, on fraternal variability; which itself is due in part to mixed ancestry, and in part to variety of nurture in the same Fraternity, both before as well as after birth. Secondly, it depends upon the variety of ancestry and nurture in different Fraternities. As to the first of the two groups of causes, they seem to affect consumptive fraternities in the same way as others, but not so with respect to the second group. The household arrangements of vigorous, of moderately vigorous, and of invalided parents are not alike. I have already spoken of infection. There is also a tradition in families that are not vigorous, of the necessity of avoiding risks and of never entering professions that involve physical hardship. There is no such tradition in families who are vigorous. Thus there must be much greater variability in the environments of a group of persons taken from the population at large, than there is in a group of consumptive families. It would be quite fair to estimate the value of n at least as low as 0·8.

We have thus three values of n to try; viz. 1, 0·9, and 0·8, of which the first is scarcely possible and the last is much the more suitable of the other two. The corresponding values of 0·73 divided by n, are + 0.73, + 0·81, and + 0·91. Referring to Table 8 we find the Grades corresponding to those deviations to be 69, 71, and 73. We should therefore expect 69, 71, or 73 per cent. of the Co-Fraternity to be non-consumptive, according to the value of n we please to adopt, and the complement to those percentages, viz. 31, 29, or 27, to be consumptive. Observation (p. 173), gave the value of 26 by one method of calculation, and of 28 by another.

Too much stress must not be laid on this coincidence, because many important points had to be slurred over, as already explained. Still, the *prima facie* result is successful, and enables us to say that so far as this evidence goes, the statistical method we have employed in treating consumptivity seems correct, and that the law of heredity found to govern all the different faculties as yet examined, appears to govern that of consumptivity also, although the constants of the formula differ slightly.

Data for Hereditary Diseases.—The knowledge of the officers of Insurance Companies as to the average value of unsound lives is by the confession of many of them far from being as exact as is desirable. [See, for example, the discussion on a memoir by G. Humphreys, Actuary to the Eagle Insurance Company, read before the Institute of Actuaries.—*Insur. Mag.* xviii. p. 178.]

Considering the enormous money value concerned, it would seem well worth the while of the higher class of those offices to combine in order to obtain a collection of completed cases for at least two generations, or better still, for three ; such as those in Examples A and B, Appendix G, but much fuller in detail. Being completed and anonymous, there could be little objection on the score of invaded privacy.   They would have no perceptible effect on the future insurances of descendants of the families, even if these were identified, and they would lay the basis of a very much better knowledge of hereditary disease than we now possess, serving as a step for fresh departures. A main point is that the cases should not be picked and chosen to support any theory, but taken as they come to hand. There must be a vast amount of good material in existence at the command of the medical officers of Insurance Companies. If it were combined and made freely accessible, it would give material for many years' work to competent statisticians, and would be certain, judging from all experience of a like kind, to lead to unexpected results.

CHAPTER XI.

Latent Elements not very numerous.—Pure Breed.—Simplification of Hereditary Inquiry.

Latent Elements not very numerous.—It is not possible that more than one half of the varieties and number of each of the parental elements, latent or personal, can on the average subsist in the offspring. For if every variety contributed its representative, each child would on the average contain actually or potentially twice the variety and twice the number of the elements (whatever they may be) that were possessed at the same stage of its life by either of its parents, four times that of any one of its grandparents, 1024 times as many as any one of its ancestors in the 10th degree, and so on, which is absurd. Therefore as regards any variety of the entire inheritance, whether it be dormant or personal, the chance of its dropping out must on the whole be equal to that of its being retained, and only one half of the varieties can on the average be passed on by inherit-

ancc. Now we have seen that the *personal* heritage
from either Parent is one quarter, therefore as the *total*
heritage is one half, it follows that the Latent Elements
must follow the same law of inheritance as the Personal
ones. In other words, either Parent must contribute
on the average only one quarter of the Latent
Elements, the remainder of them dropping out and
their breed becoming absolutely extinguished.

There seems to be much confusion in current ideas
about the extent to which ancestral qualities are
transmitted, supposing that what occurs occasionally
must occur invariably. If a maternal grandparent be
found to contribute some particular quality in one
case, and a paternal grandparent in another, it seems
to be argued that both contribute elements in every
case. This is not a fair inference, as will be seen by
the following illustration. A pack of playing cards
consists, as we know, of 13 cards of each sort—hearts,
diamonds, spades, and clubs. Let these be shuffled
together and a batch of 13 cards dealt out from them,
forming the deal, No. 1. There is not a single card
in the entire pack that may not appear in these 13,
but assuredly they do not all appear. Again, let the
13 cards derived from the above pack, which we will
suppose to have green backs, be shuffled with another
13 similarly obtained from a pack with blue backs,
and that a deal, No. 2, of 13 cards be made from the
combined batches. The result will be of the same kind
as before. Any card of either of the two original
packs may be found in the deal, No. 2, but certainly

not all of them. So I conceive it to be with hereditary transmission. No given pair can possibly transmit the whole of their ancestral qualities; on the other hand, there is probably no description of ancestor whose qualities have not been in some cases transmitted to a descendant.

The fact that certain ancestral forms persist in breaking out, such as the zebra-looking stripes on the donkey, is no argument against this view. The reversion may fairly be ascribed to precisely the same cause that makes it almost impossible to wholly destroy the breed of certain weeds in a garden, inasmuch as they are prolific and very hardy, and wage successful battle with their vegetable competitors whenever they are not heavily outmatched in numbers.

If the Personal and Latent Elements are transmitted on the average in equal numbers, it is difficult to suppose that there can be much difference in their variety.

Pure Breed.—In a perfectly pure breed, maintained during an indefinitely long period by careful selection, the tendency to regress towards the M of the general population, would disappear, so far as that tendency may be due to the inheritance of mediocre ancestral qualities, and not to causes connected with the relative stability of different types. The Q of Fraternal Deviations from their respective true Mid-Fraternities which we called *b*, would also diminish, because it is partly dependent on the children in the same family taking

variously after different and unlike progenitors. But
the difference between b in a mixed breed such as we
have been considering, and the value which we may
call β, which it would have in a pure breed, would be
very small. . Suppose the Prob: Error of the implied
Stature of each separate Grand-Parent to be even as
great as the Q of the general Population, which is 1·7
inch (it would be less, but we need not stop to discuss
its precise value), then the Prob : Error of the implied
Mid-Grand-Parental stature would be $\sqrt{\frac{1}{4}} \times 1·7$ inch, or
say 0·8 inch. The share of this, which would on the
average be transmitted to the child, would be only $\frac{1}{4}$ as
much, or 0·2. From all the higher Ancestry, put
together, the contribution would be much less even than
this small value, and we may disregard it. It results
that b^2 is a trifle greater than $\beta^2 + 0·04$. But $b = 1·0$;
therefore β is only a trifle less than 0·98.

Simplification of Hereditary Inquiry. — These
considerations make it probable that inquiries into
human heredity may be much simplified. They assure
us that the possibilities of inheritance are not likely to
differ much more than the varieties actually observed
among the members of a large Fraternity. If then we
have full life-histories of the Parents and of numerous
Uncles and Aunts on both sides, we ought to have a
very fair basis for hereditary inquiry. Information of
this limited kind is incomparably more easy to obtain
than that which I have hitherto striven for, namely,
family histories during four successive generations.

When the "children" in the pedigree are from 40 to
55 years of age, their own life-histories are sufficiently
advanced to be useful, though they are incomplete,
and it is still easy for them to compile good histories
of their Parents, Uncles, and Aunts. Friends who
knew them all would still be alive, and numerous
documents such as near relations or personal friends
preserve, but which are mostly destroyed at their
decease, would still exist. If I were undertaking
a fresh inquiry in order to verify and to extend my
previous work, it would be on this basis. I should not
care to deal with any family that did not number at
least six adult children, and the same number of uncles
and aunts on both the paternal and maternal sides.
Whatever could be learnt about the grandparents
and their brothers and sisters, would of course be
acceptable, as throwing further light. I should however expect that the peculiarities distributed among
any large Fraternity of Uncles and Aunts would fairly
indicate the variety of the Latent Elements in the
Parent. The complete heritage of the child, on the
average of many cases, might then be assigned as
follows: One quarter to the personal characteristics of
the Father; one quarter to the average of the personal
characteristics of the Fraternity taken as a whole, of
whom the Father was one of the members; and similarly
as regards the Mother's side.

CHAPTER XII.

SUMMARY

THE investigation now concluded is based on the fact that the characteristics of any population that is in harmony with its environment, may remain statistically identical during successive generations. This is true for every characteristic whether it be affected to a great degree by a natural selection, or only so slightly as to be practically independent of it. It was easy to see in a vague way, that an equation admits of being based on this fact; that the equation might serve to suggest a theory of descent, and that no theory of descent that failed to satisfy it could possibly be true.

A large part of the book is occupied with preparations for putting this equation into a working form. Obstacles in the way of doing so, which I need not recapitulate, appeared on every side; they had to be confronted in turns, and then to be either evaded or overcome. The final result was that the higher methods of statistics, which consist in applications of the law of Frequency of Error, were found eminently suitable for expressing the processes of heredity. By their aid, the desired equation was thrown into an exceedingly simple form of approximative accuracy, and it became easy to compare both it and its consequences with the varied results of observation, and thence to deduce numerical results.

A brief account of the chief hereditary processes occupies the first part of the book. It was inserted principally in order to show that a reasonable *a priori* probability existed, of the law of Frequency of Error being found to apply to them. It was not necessary for that purpose to embarrass ourselves with any details of theories of heredity beyond the fact, that descent either was particulate or acted as if it were so. I need hardly say that the idea, though not the phrase of particulate inheritance, is borrowed from Darwin's provisional theory of Pangenesis, but there is no need in the present inquiry to borrow more from it. Neither is it requisite to take Weissmann's views into account, unless I am mistaken as to their scope. It is freely conceded that particulate inheritance is not the only factor to be reckoned with in a complete theory of heredity, but that the stability of the organism has also to be regarded. This may perhaps become a factor of great importance in forecasting the issue of highly bred animals, but it was not found to exercise any sensible influence on those calculations with which this book is chiefly concerned. Its existence has therefore been only noted, and not otherwise taken into account.

The data on which the results mainly depend had to be

collected specially, as no suitable material for the purpose
was, so far as I know, in existence.   This was done by
means of an offer of prizes some years since, that placed
in my hands a collection of about 160 useful Family
Records.   These furnished an adequate though only
just an adequate supply of the required data.   In order
to show the degree of dependence that might be placed
on them they were subjected to various analyses, and
the result proved to be even more satisfactory than
might have been fairly hoped for.   Moreover the errors
in the Records probably affect different generations in
the same way, and would thus be eliminated from the
comparative results.

As soon as the character of the problem of Filial descent
had become well understood, it was seen that a general
equation of the same form as that by which it was
expressed, also expressed the connection between Kinsmen in every degree.   The unexpected law of universal
Regression became a theoretical necessity, and on
appealing to fact its existence was found to be conspicuous.   If the word "peculiarity" be used to signify
the difference between the amount of any faculty possessed by a man, and the average of that possessed
by the population at large, then the law of Regression
may be described as follows.   Each peculiarity in a man
is shared by his kinsmen, but *on the average* in a less
degree.   It is reduced to a definite fraction of its
amount, quite independently of what its amount might
be.   The fraction differs in different orders of kinship,
becoming smaller as they are more remote.   When the

kinship is so distant that its effects are not worth taking
into account, the peculiarity of the man, however remarkable it may have been, is reduced to zero in his
kinsmen. This apparent paradox is fundamentally due
to the greater frequency of mediocre deviations than of
extreme ones, occurring between limits separated by
equal widths.

Two causes affect family resemblance; the one is
Heredity, the other is Circumstance. That which is
transmitted is only a sample taken partly through the
operation of "accidents," out of a store of otherwise unused material, and circumstance must always play a
large part in the selection of the sample. Circumstance
comprises all the additional accidents, and all the peculiarities of nurture both before and after birth, and every
influence that may conduce to make the characteristics
of one brother differ from those of another. The
circumstances of nurture are more varied in Co-Fraternities than in Fraternities, and the Grandparents
and previous ancestry of members of Co-Fraternities
differ; consequently Co-Fraternals differ among themselves more widely than Fraternals.

The average contributions of each separate ancestor
to the heritage of the child were determined apparently
within narrow limits, for a couple of generations at
least. The results proved to be very simple; they
assign an average of one quarter from each parent,
and one sixteenth from each grandparent. According
to this geometrical scale continued indefinitely backwards, the total heritage of the child would be

accounted for, but the factor of stability of type has to be reckoned with, and this has not yet been adequately discussed.

The ratio of filial Regression is found to be so bound up with co-fraternal variability, that when either is given the other can be calculated. There are no means of deducing the measure of fraternal variability solely from that of co-fraternal. They differ by an element of which the value is thus far unknown. Consequently the measure of fraternal variability has to be calculated separately, and this cannot be done directly, owing to the small size of human families. Four different and indirect methods of attacking the problem suggested themselves, but the calculations were of too delicate a kind to justify reliance on the R.F.F. data. Separate and more accurate measures, suitable for the purpose, had therefore to be collected. The four problems were then solved by their means, and although different groups of these measures had to be used with the different problems, the results were found to agree together.

The problem of expressing the relative nearness of different degrees of kinship, down to the point where kinship is so distant as not to be worth taking into account, was easily solved. It is merely a question of the amount of the Regression that is appropriate to the different degrees of kinship. This admits of being directly observed when a sufficiency of data are accessible, or else of being calculated from the values found in this inquiry. A table of these Regressions was given.

Finally, considerations were offered to show that latent elements probably follow the same law as personal ones, and that though a child may inherit qualities from any one of his ancestors (in one case from this one, and in another case from another), it does not follow that the store of hidden property so to speak, that exists in any parent, is made up of contributions from all or even very many of his ancestry.

Two other topics may be mentioned. Reason was given in p. 16 why experimenters upon the transmission of Acquired Faculty should not be discouraged on meeting with no affirmative evidence of its existence in the first generation, because it is among the grandchildren rather than among the children that it should be looked for. Again, it is hardly to be expected that an acquired faculty, if transmissible at all, would be transmitted without dilution. It could at the best be no more than a variation liable to Regression, which would probably so much diminish its original amount on passing to the grandchildren as to render it barely recognizable. The difficulty of devising experiments on the transmission of acquired faculties is much increased by these considerations.

The other subject to be alluded to is the fundamental distinction that may exist between two couples whose personal faculties are naturally alike. If one of the couples consist of two gifted members of a poor stock, and the other of two ordinary members of a gifted stock, the difference between them will betray itself in their offspring. The children of the former will tend to regress; those of the latter will not. The value of a good stock to the well-being of future generations is therefore obvious, and it is well to recall attention to an early sign by which we may be assured that a new and gifted variety possesses the necessary stability to easily originate a new stock. It is its refusal to blend freely with other forms. Some among the members of the same fraternity might possess the characteristics in question with much completeness, and the remainder hardly or not at all. If this alternative tendency was also witnessed among cousins, there could be little doubt that the new variety was of a stable character, and therefore capable of being easily developed by interbreeding into a pure and durable race.

TABLES.

TABLE 1.

Strength of Pull.	No. of cases observed.	Percentages.	
		No. of cases observed.	Sums from beginning.
Under 50 lbs.	10	2	2
,, 60 ,,	42	8	10
,, 70 ,,	140	27	37
,, 80 ,,	168	33	70
,, 90 ,,	113	21	91
,, 100 ,,	22	4	95
Above 100 ,,	24	5	100
Total	519	100	

STRENGTH OF PULL.
519 Males aged 23–26.
From measures made at the International Health Exhibition in 1884.

TABLE 2.

DATA FOR SCHEMES OF DISTRIBUTION of various qualities and faculties among the persons measured at the Anthropometric Laboratory in the International Exhibition of 1884.

Subject of measurement	Age	Unit of measurement	Sex	No. of persons in the group.	Values at the undermentioned Grades, from 0° to 160°.										
					5°	10°	20°	30°	40°	50°	60°	70°	80°	90°	95°
Height, standing, without shoes .	23–51	Inches	M.	811	63·2	64·5	65·8	66·5	67·3	67·9	68·5	69·2	70·0	71·3	72·4
			F.	770	58·9	59·9	61·3	62·1	62·7	63·3	63·9	64·6	65·3	66·4	67·3
Height, sitting, from seat of chair .	23–51	Inches	M.	1013	33·6	34·2	34·9	35·3	35·4	36·0	36·3	36·7	37·1	37·7	38·2
			F.	775	31·8	32·3	32·9	33·3	33·6	33·9	34·2	34·6	34·9	35·6	36·0
Span of arms . . .	23–51	Inches	M.	811	65·0	66·1	67·2	68·2	69·0	69·9	70·6	71·4	72·3	73·6	74·8
			F.	770	58·6	59·5	60·7	61·7	62·4	63·0	63·7	64·5	65·4	66·7	68·0
Weight in ordinary indoor clothes .	23–26	Pounds	M.	520	121	125	131	135	139	143	147	150	156	165	172
			F.	276	102	105	110	114	118	122	129	132	136	142	149
Breathing capacity.	23–26	Cubic Inches	M.	212	161	177	187	199	211	219	226	236	248	277	290
			F.	277	92	102	115	124	131	138	144	151	164	177	186
Strength of pull as archer with bow.	23–26	Pounds	M.	519	56	60	64	68	71	74	77	80	82	89	96
			F.	276	30	32	34	36	38	40	42	44	47	51	54
Strength of squeeze with strongest hand	23–26	Pounds	M.	519	67	71	76	79	82	85	88	91	95	100	104
			F.	276	36	39	43	47	49	52	55	58	62	67	72
Swiftness of blow .	23–26	Ft. per second	M.	516	13·2	14·1	15·2	16·2	17·3	18·1	19·1	20·0	20·9	22·3	23·6
			F.	271	9·2	10·1	11·3	12·1	12·8	13·4	14·0	14·5	15·1	16·3	16·9
Sight, keenness of — by distance of reading diamond test-type .	23–26	Inches	M.	398	13	17	20·	22	23	25	26	28	30	32	34
			F.	433	10	12	16	19	22	24	26	27	29	31	32

TABLE 3.

Deviations from **M** in each of the series in Table 2, after reduction to a Scale in which Q' = 1, where Q' is the *Mean* of the observed Deviations at the Grades 20°, 30°, 70°, and 80°.

Subject of measurement.	Values of Q'	Unit of measurement in Table 2.	Sex	No. of persons	Deviations reckoned in units of Q'.										
					5°	10°	20°	30°	40°	50°	60°	70°	80°	90°	95°
Height, standing, without shoes	1·72	Inches	M.	811	2·73	1·98	1·22	0·81	0·35	0	0·35	·76	1·22	1·98	2·61
	1·62		F.	770	2·71	2·10	1·23	·74	·37	0	·37	·80	1·23	1·91	2·46
Height, sitting, from seat of chair	0·95	Inches	M.	1013	2·52	1·89	1·15	·73	·63	0	·31	·73	1·15	1·79	2·31
	0·82		F.	775	2·55	1·95	1·22	·73	·36	0	·36	·85	1·22	2·07	2·55
Span of arms	2·07	Inches	M.	811	2·36	1·83	1·30	·82	·43	0	·33	·72	1·16	1·79	2·36
	1·87		F.	770	2·35	1·87	1·23	·69	·32	0	·37	·80	1·28	1·98	2·67
Weight in ordinary indoor clothes	10·00	Pounds	M.	520	2·20	1·80	1·20	·80	·40	0	·40	·70	1·30	2·20	2·90
	11·00		F.	276	1·80	1·60	1·10	·70	·40	0	·60	·90	1·30	1·80	2·40
Breathing capacity	24·50	Cubic Inches	M.	212	2·32	1·68	1·28	·80	·32	0	·28	·68	1·16	2·82	2·84
	19·00		F.	277	2·39	1·87	1·20	·73	·36	0	·31	·67	1·35	2·03	2·49
Strength of pull as archer with bow	7·50	Pounds	M.	519	2·39	1·86	1·33	·80	·40	0	·40	·80	1·06	1·99	2·92
	5·22		F.	276	1·92	1·06	·80	·53	·27	0	·27	·53	·93	1·46	1·86
Strength of squeeze with strongest hand	7·75	Pounds	M.	519	2·32	1·81	1·16	·77	·39	0	·39	·77	1·29	1·93	2·45
	7·50		F.	276	2·12	1·73	1·20	·66	·40	0	·40	·80	1·33	1·99	2·66
Swiftness of blow	2·37	Ft. per second	M.	516	2·06	1·68	1·22	·80	·34	0	·42	·80	1·18	1·77	2·31
	1·55		F.	271	2·71	2·13	1·35	·84	·38	0	·38	·71	1·10	1·87	2·26
Sight, keenness of — by distance of reading diamond test-type	4·00	Inches	M.	398	3·00	2·00	1·25	·75	·50	0	·25	·75	1·25	1·75	2·25
	5·22		F.	433	2·66	2·28	1·52	·95	·38	0	·38	·57	·95	1·33	1·52
SUMS					43·11	33·12	21·96	13·65	7·00	0	6·57	13·34	21·46	33·96	43·82
MEANS					2·40	1·84	1·22	0·76	0·39	0	0·37	0·74	1·19	1·89	2·43
					2·44	1·87	1·24	0·77	0·40	0	0·38	0·75	1·21	1·92	2·47
MEANS multiplied by 1·015, to change unit to Q'=1															
Normal Values, when Q' = 1					2·44	1·90	1·25	0·78	0·38	0	0·38	0·78	1·25	1·90	2·44

Tables 4 to 8 inclusive give data for drawing Normal Curves of Frequency and Distribution. They also show the way in which the latter is derived from the values of the Probability Integral.

The equation for the Probablity Curve[1] is $y = k\,e^{-h^2x^2}$ in which h is "the Measure of Precision." By taking k and h each as unity, the values in Table 4 are computed.

TABLE 4.

Data for a Normal Curve of Frequency.

$$y = e^{-x^2}$$

x	y	x	y	x	y	x	y
0	1·00	± 1·0	0·37	± 2·0	0·0183	± 3·0	0·0001
± 0·2	0·96	± 1·2	0·23	± 2·2	0·0079		
± 0·4	0·85	± 1·4	0·14	± 2·4	0·0032	± infi-	0·0000
± 0·6	0·70	± 1·6	0·78	± 2·6	0·0012	nity	
± 0·8	0·53	± 1·8	0·40	± 2·8	0·0004		

TABLE 5.

Values of the Probability Integral, $\dfrac{2}{\sqrt{\pi}}\displaystyle\int_0^t e^{-t^2}\,dt$, for Argument t.

$t\ (=hx)$	·0	·1	·2	·3	·4	·5	·6	·7	·8	·9
0	0·00	0·11	0·22	0·33	0·43	0·52	0·60	0·68	0·74	0·80
1·0	0·843	0·880	0 910	0·934	0·952	0·966	0·976	0·984	0·989	0·923
2·0	·9953	·9970	·9981	·9989	·9993	·9996	·9998	9999	·9999	·9999
infinite	1·0000									

When $t = ·4769$ the corresponding tabular entry would be ·50; therefore, ·4769 is the value of the "Probable Error."

[1] See Merriman *On the Method of Least Squares* (Macmillan, 1885), pp. 26, 186, where fuller Tables than 4, 5, and 6 will be found

TABLE 6.

Values of the Probability Integral for Argument $\frac{t}{0.4769}$; that is, when the unit of measurement = the Probable error.

Multiples of the Probable Error.	·0	·1	·2	·3	·4	·5	·6	·7	·8	·9
0	0·00	0·65	0·11	0·16	0·21	0·26	0·31	0·36	0·41	0·46
1·0	·50	·54	·58	·62	·66	·69	·72	·75	·78	·80
2·0	·82	·84	·86	·88	·89	·91	·92	·93	·94	·95
3·0	·957	·964	·969	·974	·978	·982	·985	·987	·990	·992
4·0	·9930	·9943	·9954	·9963	·9970	·9976	·9981	·9985	·9988	·9990
5·0	·9993	·9994	·9996	·9997	·9997	·9998	·9998	·9999	·9999	·9999
infinite	1·000									

Tables 5 and 6 show the proportion of cases in any Normal system, in which the amount of Error lies within various extreme values, the total number of cases being reckoned as 1·0. Here no regard is paid to the sign of the Error, whether it be *plus* or *minus*, but its amount is alone considered. The unit of the scale by which the Errors are measured, differs in the two Tables. In Table 5 it is the "Modulus," and the result is that the Errors in one half of the cases, that is in 0·50 of them lie within the extreme value (found by interpolation) of 0·4769, while the other half exceed that value. In Table 6 the unit of the scale is 0·4769. It is derived from Table 5 by dividing all the tabular entries by that amount. Consequently one half of the cases have Errors that do not exceed 1·0 in terms of the new unit, and that unit is the Probable Error of the System. It will be seen in Table 6 that the entry of ·50 stands opposite to the argument of 1·0.

If it be desired to transform Tables 5 and 6 into others that shall show the proportion of cases in which the *plus* Errors and the *minus* Errors respectively lie within various extreme limits, their entries would have to be halved.

Let us suppose this to have been done to Table 6, and that a new Table, which it is not necessary to print, has been thereby produced and which we will call 6a. Next multiply all the entries in the new Table by 100 in order to make them refer to a total number of 100 cases, and call this second Table 6b. Lastly make a converse Table to 6b; one in which the arguments of 6b become the entries, and the entries of 6b become the arguments. From this the Table 7

is made. For example, in Table 6, opposite to the argument 1·00, the
entry of ·50 is found ; that entry becomes ·25 in 6a, and 25 in 6b.
In Table 7 the argument is 25, and the corresponding entry is 1·00.
The meaning of this is, that in 25 per cent. of the cases the greatest
of the Errors just attains to ± 1·0. Similarly Table 7 shows that
the greatest of the Errors in 30 per cent. of the cases, just attains
to ± 1·25 ; in 40 per cent. to 1·90, and so on. These various percentages correspond to the centesimal Grades in a Curve of Distribution, when the Grade 0° is placed at the middle of the axis, which
is the point where it is cut by the Curve, and where the other
Grades are reckoned outwards on either hand, up to + 50° on the
one side, and to — 50° on the other.

To recapitulate :—In order to obtain Table 7 from the primary
Table 5, we have to halve each of the entries in the body of Table 5,
then to multiply each of the arguments by 100, and divide it by
·4769. Then we expand the Table by interpolations, so as to
include among its entries every whole number from 1 to 99 inclusive.
Selecting these and disregarding the rest, we turn them into the
arguments of Table 7, and we turn their corresponding arguments
into the entries in Table 7.

TABLE 7.

ORDINATES TO NORMAL CURVE OF DISTRIBUTION

on a scale whose unit = the Probable Error ; and in which the 100 Grades run
from 0° to +50° on the one side, and to - 50° on the other.

Grades.	0	1	2	3	4	5	6	7	8	9
0	0·00	0·04	0·07	0·11	0·15	0.19	0·22	0·26	0·30	0·34
10	0·38	0·41	0·45	0·49	0·53	0·57	0·61	0·65	0·69	0·74
20	0·78	0·82	0·86	0·97	0·95	1·00	1·05	1·10	1·15	1·20
30	1·25	1·30	1·36	1·42	1·47	1·54	1·60	1·67	1·74	1·82
40	1·90	1·99	2·08	2·19	2·31	2·44	2·60	2·79	3·05	3·45

But in the Schemes, the 100 Grades do not run from—50° through
0° to + 50°, but from 0° to 100°. It is therefore convenient to
modify Table 7 in a manner that will admit of its being used
directly for drawing Schemes without troublesome additions or
subtractions. This is done in Table 8, where the values from
50° onwards, and those from 50° backwards are identical with
those in Table 7 from 0° to ± 50°, but the first half of those
in Table 8 are positive and the latter half are. negative.

TABLE 8.

ORDINATES TO NORMAL CURVE OF DISTRIBUTION on a scale whose unit = the Probable Error, and in which the 100 Grades run from 0° to 100°.

Grades	0	1	2	3	4	5	6	7	8	9
0	-∞	-3·45	-3·05	-2·79	-2·60	-2·44	-2·31	-2·19	-2·08	-1·99
10	-1·90	-1·82	-1·74	-1·67	-1·60	-1·54	-1·47	-1·42	-1·36	-1·30
20	-1·25	-1·20	-1·15	-1·10	-1·05	-1·00	-0·95	-0·91	-0·86	-0·82
30	-0·78	-0·74	-0·69	-0·65	-0·61	-0·57	-0·53	-0·49	-0·45	-0·41
40	-0·38	-0·34	-0·30	-0·26	-0·22	-0·19	-0·15	-0·11	-0·07	-0·04
50	0·00	+0·04	+0·07	+0·11	+0·15	+0·19	+0·22	+0·26	+0·30	+0·34
60	+0·38	+0·41	+0·45	+0·49	+0·53	+0·57	+0·61	+0·65	+0·69	+0·74
70	+0·78	+0·82	+0·86	+0·91	+0·95	+1·00	+1·05	+1·10	+1·15	+1·20
80	+1·25	+1·30	+1·36	+1·42	+1·47	+1·54	+1·60	+1·67	+1·74	+1·82
90	+1·90	+1·99	+2·08	+2·19	+2·31	+2·44	+2·60	+2·79	+3·05	+3·45

Examples of the way in which Table 8 is to be read :—

The ordinate at 0° is − infinity; at 10° it is − 1·90; at 11° it is − 1·82; at 25° it is − 1·00; at 75° it is + 1·00. The Table does not go beyond Grade 99°, where the ordinate is + 3·45. At the Grade 100°, the ordinate would be + infinity.

TABLE 9.

MARRIAGE SELECTION IN RESPECT TO STATURE.

The 205 male parents and the 205 female parents are each divided into three groups—T, M, and S, and *t*, *m*, and *s*, respectively—that is, Tall, Medium, and Short (medium male measurements being taken as 67 inches, and upwards to 70 inches). The number of marriages in each possible combination between them were then counted, with the result that men and women of contrasted heights, Short and Tall, or Tall and Short, married about as frequently as men and women of similar heights, both Tall or both Short; there were 32 cases of the one to 27 of the other.

S., t. 12 cases.	M., t. 20 cases.	T., t. 18 cases.
S., m. 25 cases.	M., m. 51 cases.	T., m. 28 cases.
S., s. 9 cases.	M., s. 28 cases.	T., s. 14 cases.

Short and tall, 12 + 14 = 32 cases.
Short and short, 9 }
Tall and tall, 18 } = 27 cases.

We may therefore regard the married folk as couples picked out of the general population at haphazard when applying the law of probabilities to heredity of stature.

TABLE 9A.

MARRIAGE SELECTION IN RESPECT TO EYE-COLOUR

in 78 Parental Couples.

Eye Colour of		No. of cases observed.	Per Cents.				Eye Colour of Husband and Wife.
Husband	Wife.		Obs.	Chance.	Observed.	Chance.	
Light	Light	29	37	37			
Hazel	Hazel	2	3	2	48	46	Alike
Dark	Dark	6	8	7			
Light	Hazel	} 18	23	15			{ Half-con-trasted
Hazel	Light				28	22	
Hazel	Dark	} 4	5	7			
Dark	Hazel						
Light	Dark	} 19	24	32	24	32	Contrasted
Dark	Light						

The chance combinations in pairs are calculated for a population containing 61·2 per cent. of Light Eye-colour, 12·7 of Hazel, and 26·1 of Dark.

TABLE 9B.

MARRIAGES OF THE ARTISTIC AND THE NOT ARTISTIC.

Rank in Pedigrees.	No. of persons.	Percentages.									
		Males.		Females.		Pairs of artistic and not artistic persons.					
						Marriages observed.			Chance combinations.		
		art.	not.	art.	not.	both art.	1 art. 1 not.	both not.	both art.	1 art. 1 not.	both not.
Parents	326	32	68	39	61	14	31	50	12	46	42
Paternal grandparents ..	280	27	73	30	70	12	31	57	8	41	51
Maternal grandparents..	288	24	76	28	72	9	41	50	7	39	54
Totals and means...	894	28	72	33	67	12	36	52	9	42	49

Tastes of Husband and Wife—alike .... $12 + 52 = 64$ $9 + 49 = 58$
„   „   „   contrasted....... 36 42

TABLE 10.

EFFECT UPON ADULT CHILDREN OF DIFFERENCES IN HEIGHT OF THEIR PARENTS.

Difference in inches between the Heights of the Parents.	Proportion per 50 of cases in which the Heights[1] of the Children deviated to various amounts from the Mid-filial Stature of their respective families.					Number of Children whose Heights were observed. (Total 525.)
	Less than 1 inch.	Less than 2 inches.	Less than 3 inches.	Less than 4 inches.	Less than 5 inches.	
Under 1 inch	21	35	43	46	48	105
1 and under 2	23	37	46	49	50	122
2 „ 3	16	34	41	45	49	112
3 „ 5	24	35	41	47	49	108
5 and above..	18	30	40	47	49	78

[1] Every female height has been transmuted to its male equivalent by multiplying it by 1·08, and only those families have been included in which the number of adult children amounted to six, at least.

NOTE.—When these figures are protracted into curves, it will be seen—(1) that they run much alike ; (2) that their peculiarities are not in sequence ; and (3) that the curve corresponding to the first line occupies a medium position. It is therefore certain that differences in the heights of the Parents have on the whole an inconsiderable effect on the heights of their Offspring.

TABLE 11 (R.F.F. Data).

NUMBER OF ADULT CHILDREN of various STATURES born of 205 MID-PARENTS of various STATURES.

(All Female Heights have been multiplied by 1·08.)

Height of the mid-parents in inches.	Heights of the adult children.														Total number of		Medians or Values of M.
	Below	62·2	63·2	64·2	65·2	66·2	67·2	68·2	69·2	70·2	71·2	72·2	73·2	Above	Adult children.	Mid-parents.	
Above 72·5...	...	...	...	...	...	...	...	...	...	...	...	1	3	...	4	5[1]	72·2
72·5...	...	...	...	...	...	...	...	1	2	1	2	7	2	4	19	6	69·9
71·5...	...	...	...	...	1	3	4	3	5	10	4	9	2	2	43	11	69·5
70·5...	1	...	1	...	1	1	3	12	18	14	7	4	3	3	68	22	68·9
69·5...	...	...	1	16	4	17	27	20	33	25	20	11	4	5	183	41	68·2
68·5...	1	...	7	11	16	25	31	34	48	21	18	4	3	...	219	49	67·6
67·5...	...	3	5	14	15	36	38	28	38	19	11	4	...	...	211	33	67·2
66·5...	...	3	3	5	8	11	14	14	13	4	2	1	...	...	78	20	66·7
65·5...	1	...	9	5	1	17	14	7	7	5	...	...	...	...	66	12	65·8
64·5...	...	1	4	4	1	5	5	1	2	...	...	...	...	...	23	5	
Below ...	2	...	2	4	1	2	2	...	1	...	...	...	...	...	14	1	
Totals	5	7	32	59	48	117	138	120	167	99	64	41	17	14	928	205	
Medians	...	...	66·3	67·8	67·9	67·7	67·9	68·3	68·5	69·0	69·0	70·0	...	...			

Note.—In calculating the medians, the entries have been taken as referring to the middle of the squares in which they stand. The reason why the headings run 62·2, 63·2, &c., instead of 62·5, 63·5, &c., is that the observations are unequally distributed between 62 and 63, 63 and 64, &c., there being a strong bias in favour of integral inches. After careful consideration, I concluded that the headings, as adopted, best satisfied the conditions. This inequality was not apparent in the case of the mid-parents.

[1] I have reprinted this Table without alteration from that published in the *Proc. Roy. Soc.,* notwithstanding a small blunder since discovered in sorting the entries between the first and second lines. It is obvious that 4 children cannot have 5 Mid-Parents. The bottom line, which looks suspicious, is correct.

TABLE 12 (R.F.F. Data).

RELATIVE NUMBER OF BROTHERS OF VARIOUS HEIGHTS TO MEN OF VARIOUS HEIGHTS, FAMILIES OF SIX BROTHERS AND UPWARDS BEING EXCLUDED.

Heights of the men in inches	Heights of their brothers in inches														Total Cases	Medians
	Below 61·7	62·2	63·2	64·2	65·2	66·2	67·2	68·2	69·2	70·2	71·2	72·2	73·2	Above 73·2		
Above 73·7	...	...	...	...	1	...	1	...	1	4	3	3	3	2	18	
73·2	1	...	...	...	...	...	1	1	2	1	3	4	...	3	16	
72·2	1	...	1	2	1	1	...	8	6	8	11	5	4	3	51	70·3
71·2	...	...	...	4	4	4	9	11	15	12	8	11	3	3	84	69·3
70·2	1	...	2	4	3	7	6	12	25	18	11	8	1	3	101	69·3
69·2	...	...	4	6	13	12	18	29	29	24	15	6	2	1	159	68·6
68·2	1	...	...	3	6	7	15	16	29	12	11	8	1	...	109	68·9
67·2	1	...	4	3	8	14	21	15	19	6	9	...	1	1	102	67·7
66·2	...	...	1	7	10	12	14	7	12	7	4	1	...	...	75	67·2
65·2	...	1	1	4	13	9	8	6	13	3	4	1	...	1	64	67·2
64·2	...	1	...	6	4	7	3	3	6	4	4	2	...	...	40	67·3
63·2	...	...	...	...	1	1	4	...	4	2	...	1	...	...	13	
62·2	...	...	...	...	...	...	1	...	...	...	...	...	...	...	1	
Below 61·7	...	...	...	...	...	...	1	1	...	1	...	1	1	...	5	
...	5	2	13	39	65	74	101	109	161	102	83	51	16	17	838	

TABLE 13 (Special Data).

RELATIVE NUMBER OF BROTHERS OF VARIOUS HEIGHTS TO MEN OF VARIOUS HEIGHTS, FAMILIES OF FIVE BROTHERS AND UPWARDS BEING EXCLUDED.

Heights of the men in inches.	Heights of their brothers in inches.													Total cases.	Medians.
	Below 63	63·5	64·5	65·5	66·5	67·5	68·5	69·5	70·5	71·5	72·5	73·5	Above 74		
74 and above	1	1	...	...	...	...	...	1	1	...	5	3	12	24	.
73·5	...	...	...	...	...	1	3	4	8	3	3	2	3	27	71·1
72·5	...	...	...	...	1	1	6	5	9	9	8	3	5	47	70·2
71·5	...	1	...	1	2	8	11	18	14	20	9	4	...	88	69·6
70·5	...	...	1	1	7	19	30	45	36	14	9	8	1	171	69·5
69·5	...	1	2	1	11	20	36	55	44	17	5	4	2	198	68·7
68·5	...	1	5	9	18	38	46	36	30	11	6	3	...	203	67·7
67·5	2	4	8	26	35	38	38	20	18	8	1	1	...	199	67·0
66·5	4	3	10	33	28	35	20	12	7	2	1	...	...	155	66·5
65·5	3	3	15	18	33	26	8	2	1	1	...	...	...	110	65·6
64·5	3	8	12	15	10	8	5	2	1	...	...	...	...	64	
63·5	5	2	8	3	3	4	1	1	...	...	...	...	1	20	
Below 63	5	5	3	3	4	2	...	...	...	...	...	...	1	23	
Totals.........	23	29	64	110	152	200	204	201	169	86	47	28	25	1329	

TABLE 14 (Special Data).

DEVIATIONS OF INDIVIDUAL BROTHERS FROM THEIR MID-FRATERNAL
STATURES.

Number of brothers in each family......	4	5.	6	7
Number of Families........................	39	23	8	6
Amount of Deviation.	Number of cases.	Number of cases.	Number of cases.	Number of cases.
Under 1 inch...............................	88	62	20	21
1 and under 2.............................	49	30	18	14
2 and under 3.............................	15	17	5	6
3 and under 4.............................	4	3	3	1
4 and above...............................		3	2	

TABLE 15.
FREQUENCY OF DIFFERENT EYE-COLOURS IN FOUR SUCCESSIVE GENERATIONS.

Sex and the No. of the (ascending) generation.	No. of cases of eye-colour observed.									Percentages.								
	1. Light blue.	2. Blue. Dark blue.	3. Grey. Blue-green.	4. Dark Grey. Hazel.	5. Light brown.	6. Brown.	7. Dark Brown.	8. Very dark brown. Black.	Totals.	1. Light blue.	2. Blue. Dark blue.	3. Grey. Blue-green.	4. Dark Grey. Hazel.	5. Light brown.	6. Brown.	7. Dark brown.	8. Very dark brown. Black.	Totals.
Males IV	13	177	136	40	2	39	44	12	463	2·8	38·2	29·4	8·6	0·4	8·4	9·5	2·6	99·9
Males III	19	234	233	84	3	79	97	24	773	2·4	30·3	30·1	10·9	0·4	10·1	12·6	3·1	99·9
Males II	30	167	236	108	8	83	74	36	742	4·0	22·5	31·8	14·6	1·1	11·2	10·0	4·8	100·0
Males I	3	89	82	47	1	37	31	9	299	1·0	28·9	27·4	15·7	0·3	12·4	10·4	3·0	100·0
General	65	687	687	279	14	238	246	81	2277	2·9	29·3	30·2	12·3	0·6	10·4	10·8	3·6	100·0
Females IV	7	182	114	48	2	70	58	19	450	1·5	29·3	25·3	10·7	0·4	15·6	12·9	4·2	99·9
Females III	22	173	241	89	7	100	98	17	742	2·9	23·3	32·5	12·1	0·9	13·5	12·5	2·3	100·0
Females II	21	210	241	98	3	78	60	24	735	2·9	28·6	32·8	13·3	0·4	10·6	8·2	3·3	100·1
Females I	6	78	82	55	5	33	22	5	286	2·1	27·3	28·7	19·2	1·7	11·5	7·7	1·7	99·0
General	56	593	678	290	17	281	233	65	2213	2·5	26·8	30·6	13·1	0·8	12·7	10·5	2·9	99·9
Males and Females IV	20	309	240	88	4	109	102	31	913	2	34	27	10	1	12	11	3	100
Males and Females III	41	407	474	173	10	179	190	41	1515	3	27	31	11	1	12	12	3	100
Males and Females II	51	377	477	206	11	161	134	60	1477	3	26	32	14	1	11	9	4	100
Males and Females I	9	167	164	102	6	70	53	14	585	1	29	28	18	1	12	9	2	100
General	121	1260	1365	569	31	519	479	146	4490	2·7	28·1	30·4	12·7	0·7	11·6	10·7	3·3	100·2

TABLE 16.

THE DESCENT OF HAZEL-EYED FAMILIES.

	Total cases.	Observed.			Percentages.		
		Light.	Hazel.	Dark.	Light.	Hazel.	Dark.
General population	4490	2746	569	1175	61·2	12·7	26·1
III. Grandparents...	449	267	61	121	60	13	27
II. Parents.................	336	165	85	86	49	25	26
I. Children..............	948	430	302	216	45	32	23

TABLE 17.

CALCULATED CONTRIBUTIONS OF EYE-COLOUR.

Contribution to the heritage from each.	Data limited to the eye-colours of the					
	2 parents.		4 grandparents.		2 parents and 4 grandparents.	
	I.		II.		III.	
	Light.	Dark.	Light.	Dark.	Light.	Dark.
Light-eyed parent.........	0·30	...	...	...	0·25	..
Hazel-eyed parent.........	0·20	0·10	...	...	0·10	0·09
Dark-eyed parent	...	0·30	...	...	...	0·25
Light-eyed grandparent..	...	...	0·16	...	0·08	...
Hazel-eyed grandparent .	...	...	0·10	0·06	0·05	0·03
Dark-eyed grandparent...	...	...	...	0·16	...	0·08
Residue, rateably assigned	0·28	0·12	0·25	0·11	0·12	0·06

TABLE 18.

EXAMPLE OF ONE CALCULATION IN EACH OF THE THREE CASES.

Ancestry and their eye-colours.	I.			II.			III.		
	No. about whom data exist.	Contribute to		No. about whom data exist.	Contribute to		No. about whom data exist.	Contribute to	
		Light.	Dark.		Light.	Dark.		Light.	Dark.
Light-eyed parents.	2	0·60	...	...	...	...	...	...	...
Hazel-eyed parents.	...	...	...	...	...	...	1	0·16	0·09
Dark-eyed parents .	...	...	...	...	...	...	1	...	0·25
Light-eyed grandparents...............	...	...	...	1	0·16	...	1	0·08	...
Hazel-eyed grandparents.......... ...	...	...	...;	2	0·20	0·12	2	0·10	0·06
Dark-eyed grandparents..	...	...	...	1	...	0·16	1	...	0·08
Residue, rateably assigned............ ...	...	0·28	0·12		0·25	0·11		0·12	0·06
Total contributions	...	0·88	0·12		0·61	0·39	.	0·46	0·54
		1·00			1·00			1·00	

TABLE 19.

OBSERVED AND CALCULATED EYE-COLOURS IN 16 GROUPS OF FAMILIES.

Those families are grouped together in whom the distribution of Light, Hazel, and Dark Eye-colour among the Parents and Grandparents is alike. Each group contains at least Twenty Brothers or Sisters.

Eye-colours of the						Total child-ren.	Number of the light eye-coloured children.			
Parents.			Grandparents.				Ob-served.	Calculated.		
Light.	Hazel.	Dark.	Light.	Hazel.	Dark.			I.	II.	III.
2	...	...	4	...	...	183	174	161	163	172
2	...	...	3	1	...	53	46	47	44	48
2	...	...	3	...	1	92	88	81	67	79
2	...	...	2	1	1	27	26	24	18	22
...	...	2	2	...	2	22	11	6	12	6
1	1	...	3	1	...	62	52	48	51	51
1	1	...	3	...	1	42	30	33	31	32
1	1	...	2	2	...	31	28	24	24	20
1	1	...	2	...	2	49	35	38	28	34
1	1	...	2	1	1	31	25	24	21	23
1	...	1	3	...	1	76	45	44	55	46
1	...	1	2	...	2	66	30	38	38	35
1	...	1	2	...	1	27	15	16	18	16
1	...	1	1	...	3	20	9	12	8	9
1	...	1	1	1	2	22	8	13	11	11
...	1	1	1	1	2	24	9	14	12	10
						629	623	601	614

TABLE 20.

OBSERVED AND CALCULATED EYE-COLOURS IN 78 SEPARATE FAMILIES, EACH OF NOT LESS THAN SIX BROTHERS OR SISTERS.

Eye-colours of the						Total children.	Number of the light eye-coloured children.			
Parents.			Grandparents.				Observed.	Calculated.		
Light.	Hazel.	Dark.	Light.	Hazel.	Dark.			I.	II.	III.
2	...	...	4	...	...	6	6	5·3	5·3	5·6
2	...	...	4	...	...	6	6	5·3	5·3	5·6
2	...	...	4	...	...	6	6	5·3	5·3	5·6
2	...	...	4	...	...	6	5	5·3	5·3	5·6
2	...	...	4	...	...	7	7	6·2	6·2	6·6
2	...	...	4	...	...	7	7	6·2	6·2	6·6
2	...	...	4	...	...	7	7	6·2	6·2	6·6
2	...	...	4	...	...	7	7	6·2	6·2	6·6
2	...	...	4	...	...	7	7	6·2	6·2	6·6
2	...	...	4	...	...	8	8	7·0	7·1	7·5
2	...	...	4	...	...	8	8	7·0	7·1	7·5
2	...	...	4	...	...	8	8	7·0	7·1	7·5
2	...	...	4	...	...	8	8	7·0	7·1	7·5
2	...	...	4	...	...	8	7	7·0	7·1	7 5
2	...	...	4	...	...	8	7	7·0	7·1	7·5
2	...	...	4	...	...	12	12	10·6	10·7	11·3
2	...	...	3	1	...	7	7	6·2	5·8	6·4
2	...	...	3	1	...	10	4	8·8	8·3	9·1
2	...	...	3	1	...	12	12	10·6	10·0	10·9
2	...	...	3	...	1	7	6	6·2	5·1	6·0
2	...	...	3	...	1	8	8	7·0	5·8	6·9
2	...	...	3	...	1	9	9	7·9	6·6	7·7
2	...	...	3	...	1	9	9	7·9	6·6	7·7
2	...	...	3	...	1	9	7	7·9	6·6	7·7
2	...	...	3	...	1	10	10	8·8	7·3	8·6
2	...	...	2	2	...	7	7	6·2	5·4	6·2
2	...	...	2	2	...	10	9	8·8	7·7	8·8
2	...	...	2	1	1	6	6	5·3	4·0	5·0
2	...	...	2	1	1	10	10	8·8	6·7	8·3
...	2	...	2	1	1	7	4	6·2	4·7	4·6
...	2	...	2	...	2	8	5	5·4	4·6	4·8
...	...	2	3	...	1	6	2	1·7	4·4	2·2
...	...	2	2	...	2	9	1	2·5	5·1	2·5
...	...	2	1	...	3	6	1	2·7	2·5	1·2
...	...	2	1	...	3	11	3	3·1	4·5	2·2
...	...	2	1	1	2	6	...	1·7	3·0	1·5
...	...	2	1	1	2	7	4	2·0	3·6	1·8
1	1	...	3	1	...	6	6	4·7	5 0	4·9
1	1	...	3	1	...	7	6	5·5	5·7	5·7
1	1	...	3	1	...	8	6	6·2	6·6	6·6
1	1	...	3	1	...	9	7	7·0	7·5	7·4
1	1	...	3	1	...	11	10	8·6	9·1	9·2

TABLE 20—*continued.*

Eye-colours of the						Total children.	Number of the light eye-coloured children.			
Parents.			Grandparents.				Observed.	Children.		
Light.	Hazel.	Dark.	Light.	Hazel.	Dark.			I.	II.	III.
1	1	...	3	...	1	9	6	7·0	6·6	6·9
1	1	...	3	...	1	11	7	8·6	8·0	8·5
1	1	...	2	2	...	7	6	5·5	5·4	4·4
1	1	...	2	2	...	9	9	7·0	6·9	5·7
1	1	...	2	2	...	11	1	8·6	8·5	6·9
1	1	...	2	...	2	6	6	4·7	3·4	4·1
1	1	...	2	...	2	6	4	4·7	3·4	4·1
1	1	...	2	...	2	8	5	6·2	4·6	5·5
1	1	...	2	...	2	9	7	7·0	5·1	6·2
1	1	...	2	1	1	6	6	4·7	4·0	4·4
1	1	...	2	1	1	10	9	7·8	6·7	7·4
1	1	...	1	3	...	9	4	7·0	5·5	6·8
1	1	...	1	1	2	8	5	6·2	4·1	5·3
1	...	1	4	...	...	7	3	4·1	6·2	4·8
1	...	1	3	...	1	6	4	3·5	4·4	3·7
1	...	1	3	...	1	7	3	4·1	5·1	4·3
1	...	1	3	...	1	8	6	4·6	5·8	4·9
1	...	1	3	...	1	8	5	4·6	5·8	4·9
1	...	1	3	...	1	8	4	4·6	5·8	4·9
1	...	1	3	...	1	9	6	5·2	6·6	5·5
1	...	1	3	...	1	9	5	5·2	6·6	5·5
1	...	1	2	...	2	6	5	3·5	3·4	3·2
1	...	1	2	...	2	6	3	3·5	3·4	3·2
1	...	1	2	...	2	8	4	4·6	4·6	4·2
1	...	1	2	...	2	10	2	5·8	5·7	5·3
1	...	1	2	...	2	14	9	8·1	8·0	7·4
1	...	1	2	1	1	7	5	4·1	4·7	4·1
1	...	1	1	2	1	7	3	4·1	4·3	3·9
1	...	1	1	1	2	7	4	4·1	3·6	3 5
1	...	1	1	...	3	8	4	4·6	3·3	3·6
1	...	1	1	...	3	8	3	4·6	3·3	3·6
1	...	1	...	1	3	6	3	3·5	2·1	2·6
...	1	1	2	...	2	6	3	4·8	3·4	2·6
...	1	1	2	1	1	9	4	7·0	6·0	4·4
...	1	1	1	...	3	13	8	10·1	5·3	4·7
...	1	1	...	4	...	7	2	5·5	4·6	3 4

TABLE 21.

ERROR IN CALCULATIONS.

Numbers of Errors of various Amounts in the 3 Calculations, Table 20, of the
Number of Light Eye-coloured Children in the 78 Families.

Data employed referring to	Amount of Errors.					Total Cases.
	0·0 to 0·5.	0·6 to 1·1	1·2 to 1·7	1·8 to 2·3	2·4 and above.	
I. The 2 parents only	19	30	18	5	6	78
II. The 4 grandparents only.....	16	28	10	10	14	78
III. The 2 parents and 4 grandparents.......................	41	17	8	4	8	78

TABLE 22.

INHERITANCE OF THE ARTISTIC FACULTY.

Parents.	Children.							
	Observed.			Per cents.				
	Number of Fraternities.	Total children.	Of whom are artistic.	Observed.		Calculated.		
				art.	not art.	art.	not art.	
Both artistic	30	148	95	64	36	60	40	
One artistic; one not..	101	520	201	39	61	39	61	
Neither artistic..... ...	150	839	173	21	79	17	83	
Totals...............	281	1507	469	100	100	100	100	

The "parents" and the "children" in this Table usually rank respectively as
Grandparents and Parents in the R.F.F. pedigrees.

APPENDIX.

A.

The following memoirs by the author, bearing on Heredity, have been variously utilised in this volume:

Experiments in Pangenesis. *Proc. Royal Soc.*, No. 127, 1871, p. 393.

Blood Relationship. *Proc. Royal Soc.*, No. 136, 1872, p. 394.

A Theory of Heredity. *Journ. Anthropol. Inst.*, 1875, p. 329.

Statistics by Intercomparison. *Phil. Mag.*, Jan. 1875.

*On the Probability of the Extinction of Families. *Journ. Anthropol. Inst.*, 1875.

Typical laws of Heredity. *Journ. Royal Inst.*, Feb. 1877.

*Geometric Mean in Vital and Social Statistics. *Proc. Royal Soc.*, No. 198, 1879. See subsequent memoir by Dr. Macalister.

Address to Anthrop. Section British Association at Aberdeen. *Journ. Brit. Assoc.*, 1885.

Regression towards Mediocrity in Hereditary Stature. *Journ. Anthropol. Inst.*, 1885.

Presidential Addresses to Anthropol. Inst., 1885, 6 and 7.

Family Likeness in Stature. *Proc. Royal Soc.*, No. 242, 1886.

Family Likeness in Eye-colour. *Proc. Royal Soc.*, No. 245, 1886.

*Good and Bad Temper in English Families. *Fortnightly Review*, July, 1887.

Pedigree Moth Breeding. *Trans. Entomolog. Soc.*, 1887. See also subsequent memoir by Mr. Merrifield, and another read by him, Dec. 1887.

Those marked with an asterisk (*) are reprinted with slight revision in the Appendices F, D, and E.

WORKS ON HEREDITY BY THE AUTHOR.
(Published by Messrs. Macmillan & Co.)

Hereditary Genius. 1869.
English Men of Science. 1874.
Inquiries into Human Faculty. 1883.

Record of Family Faculties.[1] 1884.   2s. 6d.
Life History Album [2] (*edited* by F. Galton). 1884. 3s. 6d. and 4s. 6d.

[1] The Record of Family Faculties consists of Tabular Forms and Directions for entering Data, with an Explanatory Preface. It is a large thin quarto book of seventy pages, bound in limp cloth. The first part of it contains a preface, with explanation of the object of the work and of the way in which it is to be used. The rest consists of blank forms, with printed questions and blank spaces to be filled with writing. The Record is designed to facilitate the orderly collection of such data as are important to a family from an hereditary point of view. It allots equal space to every direct ancestor in the nearer degrees, and is supposed to be filled up in most cases by a parent, say the father of a growing family. If he takes pains to make inquiries of elderly relatives and friends, and to seek in registers, he will be able to ascertain most of the required particulars concerning not only his own parents, but also concerning his four grandparents ; and he can ascertain like particulars concerning those of his wife.   Therefore his children will be provided with a large store of information about their two parents, four grandparents, and eight great-grandparents, which form the whole of their fourteen nearest ancestors. A separate schedule is allotted to each of them. Space is afterwards provided for the more important data concerning many at least, of the brothers and sisters of each direct ancestor. The schedules are followed by Summary Tables, in which the distribution of any characteristic throughout the family at large may be compendiously exhibited.

[2] The Life History Album was prepared by a Sub-Committee of the Collective Investigation Committee of the British Medical Association. It is designed to serve as a continuous register of the principal biological facts in the life of its owner. The book begins with a few pages of explanatory remarks, followed by tables and charts. The first table is to contain a brief medical history of each member of the near ancestry of the owner. This is followed by printed forms on which the main facts of the owner's growth and development from birth onwards may be registered, and by charts on which measurements may be laid down at appropriate intervals and compared with the curves of normal growth. Most of the required data are such as any intelligent person is capable of recording ; those that refer to illnesses should be brief and technical, and ought to be filled up by the medical attendant. Explanations are given of the most convenient tests of muscular force, of keenness of eyesight and hearing, and of the colour sense. The 4s. 6d. edition contains a card of variously coloured wools to test the sense of colour.

₊ These two works pursue similar objects of personal and scientific utility, along different paths. The Album is designed to lay the foundation of a practice

of maintaining trustworthy life-histories that shall be of medical service in after-life to the person who keeps them. The Record shows how the life histories of members of the same family may be collated and used to forecast the development in mind and body of the younger generation of that family. Both works are intended to promote the registration of a large amount of information that has hitherto been allowed to run to waste in oblivion, instead of accumulating and forming stores of recorded experience for future personal use, and from which future inquirers into heredity may hope to draw copious supplies.

B.

PROBLEMS BY J. D. HAMILTON DICKSON, FELLOW AND TUTOR OF
ST. PETER'S COLLEGE, CAMBRIDGE.

(*Reprinted from Proc. Royal Soc.*, No. 242, 1886, *p.* 63.)

Problem 1.—A point P is capable of moving along a straight line P'OP, making an angle $\tan^{-1}\frac{1}{3}$ with the axis of y, which is drawn through O the mean position of P; the probable error of the projection of P on Oy is 1·22 inch: another point p, whose mean position at any time is P, is capable of moving from P parallel to the axis of x (rectangular co-ordinates) with a probable error of 1·50 inch. To discuss the "surface of frequency" of p.

1. Expressing the "surface of frequency" by an equation in x, y, z, the exponent, with its sign changed, of the exponential which appears in the value of z in the equation of the surface is, save as to a factor,

$$\frac{y^2}{(1·22)^2} + \frac{(3x-2y)^2}{9(1·50)^2} \quad \cdots \cdots \quad (2)$$

hence all sections of the "surface of frequency" by planes parallel to the plane of xy are ellipses, whose equations may be written in the form,

$$\frac{y}{(1·22)^2} + \frac{(3x-2y)^2}{9(1·50)^2} = C, \text{ a constant} \cdots \quad (2)$$

2. Tangents to these ellipses parallel to the axis of y are found,

by differentiating (2) and putting the coefficient of dy equal to zero, to meet the ellipses on the line,

$$\left. \begin{array}{c} \dfrac{y}{(1\cdot22)^2} - 2\dfrac{3x-2y}{9(1\cdot50)^2} = 0, \\[3mm] \text{that is} \qquad \dfrac{y}{x} = \dfrac{\dfrac{6}{9(1\cdot50)^2}}{\dfrac{1}{(1\cdot22)^2} + \dfrac{4}{9(1\cdot50)^2}} = \dfrac{6}{17\cdot6} \end{array} \right\} \quad \cdots \cdots \quad (3)$$

or, approximately, on the line $y = \frac{1}{3}x$. Let this be the line OM. (See Fig. 11, p. 101.)

From the nature of conjugate diameters, and because P is the mean position of p, it is evident that tangents to these ellipses parallel to the axis of x meet them on the line $x = \frac{2}{3}y$, viz., on OP.

3. Sections of the "surface of frequency" parallel to the plane of xz, are, from the nature of the question, evidently curves of frequency with a probable error $1\cdot50$, and the locus of their vertices lies in the plane z OP.

Sections of the same surface parallel to the plane of yz are got from the exponential factor (1) by making x constant. The result is simplified by taking the origin on the line OM. Thus putting $x = x_1$ and $y = y_1 + y'$, where by (3)

$$\frac{y_1}{(1\cdot22)^2} = 2\frac{3x_1 - 2y_1}{9(1\cdot50)^2} = 0$$

the exponential takes the form

$$\left\{ \frac{1}{(1\cdot22)^2} + \frac{4}{9(1\cdot50)^2} \right\} y'^2 + \left\{ \frac{y_1^2}{(1\cdot22)^2} + \frac{(3x_1 - 2y_1)^2}{9(1\cdot50)^2} \right\} \quad \cdots \quad (4)$$

whence, if e be the probable error of this section,

$$\left. \begin{array}{c} \dfrac{1}{e^2} = \dfrac{1}{(1\cdot22)^2} + \dfrac{4}{9(1\cdot50)^2} \\[3mm] \text{or [on referring to (3)]} \quad e = 1\cdot50 \ \sqrt{\dfrac{9}{17\cdot6}} \end{array} \right\} \quad \cdots \cdots \quad (5)$$

that is, the probable error of sections parallel to the plane of yz is nearly $\dfrac{1}{\sqrt{2}}$ times that of those parallel to the plane of xz, and the locus of their vertices lies in the plane zOM.

It is important to notice that all sections parallel to the same co-ordinate plane have the same probable error.

4. The ellipses (2) when referred to their principal axes become, after some arithmetical simplification,

$$\frac{x'^2}{20\cdot68} + \frac{y'^2}{5\cdot92} = \text{constant,} \quad \ldots \quad (6)$$

the major axis being inclined to the axis of x at an angle whose tangent is $0\cdot5014$. [In the approximate case the ellipses are $\frac{x'^2}{7} + \frac{y^2}{2} = \text{const.}$, and the major axis is inclined to the axis of x at an angle $\tan^{-1}\frac{1}{2}$.]

5. The question may be solved in general terms by putting $\text{YON} = \theta$, $\text{XOM} = \phi$, and replacing the probable errors $1\cdot22$ and $1\cdot50$ by a and b respectively; then the ellipses (2) are,

$$\frac{y^2}{a^2} + \frac{(x - y\tan\theta)^2}{b^2} = \text{C.} \quad \ldots \quad (7)$$

equation (3) becomes

$$\left.\begin{array}{c} \dfrac{y^2}{a^2} + \tan\theta\dfrac{x - y\tan\theta}{b^2} = 0 \\[3mm] \dfrac{y}{x} - \tan\phi = \dfrac{a^2\tan\theta}{b^2 + a^2\tan^2\theta} \end{array}\right\} \quad \ldots \quad (8)$$

or

and (5) becomes

$$\frac{1}{e^2} = \frac{1}{a^2} + \frac{\tan^2\theta}{b^2} \quad \ldots \quad (9)$$

whence

$$\frac{\tan\phi}{\tan\theta} = \frac{e^2}{b^2} \quad \ldots \quad (10)$$

If c be the probable error of the projection of p's whole motion on the plane of xz, then

$$c^2 = a^2\tan^2\theta + b^2,$$

which is independent of the distance of p's line of motion from the axis of x. Hence also

$$\frac{\tan\phi}{\tan\theta} = \frac{a^2}{b^2} \quad \ldots \quad (11)$$

Problem 2.—An index q moves under some restraint up and down a bar AQB, its mean position for any given position of the bar being Q ; the bar, always carrying the index with it, moves under some restraint up and down a fixed frame YMY', the mean position of Q being M : the movements of the index relatively to the bar and of the bar relatively to the frame being quite independent. For any given observed position of q, required the most probable position of Q (which cannot be observed); it being known that the probable error of q relatively to Q in all positions is b, and that of Q relatively to M is c. The ordinary law of error is to be assumed.

If in any one observation, MQ $= x$, Q$q = y$, then the law of error requires

$$\frac{x^2}{c^2} + \frac{y^2}{b^2} \qquad \qquad (12)$$

to be a minimum, subject to the condition

$$x + y = a, \text{ a constant.}$$

Hence we have at once, to determine the most probable values of x', y',

$$\frac{x'}{c^2} = \frac{y'}{b^2} = \frac{a}{b^2 + c^2} \qquad \qquad (13)$$

and the most probable position of Q, measured from M, when q's observed distance from M is a, is

$$\frac{c^2}{b^2 + c^2}\, a.$$

It also follows at once that the probable error v of Q (which may be obtained by substituting $a - x$ for y in (12)) is given by

$$\frac{1}{v^2} = \frac{1}{c^2} + \frac{1}{b^2}, \text{ or } v = \frac{bc}{\sqrt{b^2 + c^2}} \qquad \qquad (14)$$

which it is important to notice, is the same for all values of a.

C.

EXPERIMENTS ON SWEET PEAS BEARING ON THE LAW OF REGRESSION.

The reason why Sweet Peas were chosen, and the methods of selecting and planting them are described in Chapter VI., p. 79. The following Table justifies their selection by the convenient and accurate method of weighing, as equivalent to that of measuring them. It will be seen that within the limits of observed variation a difference of 0·172 grain in weight corresponds closely to an average difference of 0·01 inch in diameter.

TABLE 1.

COMPARISON OF WEIGHTS OF SWEET PEAS WITH THEIR DIAMETERS.

Distinguishing letter of seed.	Weight of one seed in grains. —— Common difference = 0·172 grain.	Length of row of 100 seeds in inches.	Diameter of one seed in hundredths of inch. Common difference = 0 01 inch.
K	1·750	21·0	21
L	1·578	20·2	20
M	1·406	19·2	19
N	1·234	17·9	18
O	1·062	17·0	17
P	·890	16·1	16
Q	·718	15·2	15

The results of the experiment are given in Table 2; its first and last columns are those that especially interest us; the remaining columns showing how these two were obtained.

It will be seen that for each increase of one unit on the part of the parent seed, there is a mean increase of only one-third of a unit in the filial seed; and again that the mean filial seed resembles the parental when the latter is about 15·5 hundredths of an inch in diameter. Taking 15·5 as the point towards which Filial Regression points, whatever may be the parental deviation from that point, the mean Filial Deviation will be in the same direction, but only one-third as much.

TABLE 2.

PARENT SEEDS AND THEIR PRODUCE.

The proportionate number of sweet peas of different sizes, produced by parent seeds also of different sizes, are given below. The measurements are those of their mean diameters, in hundredths of an inch.

Diameter of Parent Seed.	Diameters of Filial Seeds.							Total.	Mean Diameter of Filial Seeds.	
	Under 15.	15-	16-	17-	18-	19-	20-	Above 21-	Observed	Smoothed
21	22	8	10	18	21	13	6	2	100	17·5	17·3
20	23	10	12	17	20	13	3	2	100	17·3	17·0
19	35	16	12	13	11	10	2	1	100	16·0	16·6
18	34	12	13	17	16	6	2	0	100	16·3	16·3
17	37	16	13	16	13	4	1	0	100	15·6	16·0
16	34	15	18	16	13	3	1	0	100	16·0	15·7
15	46	14	9	11	14	4	2	0	100	15·3	15·4

This point is so low in the scale, that I possess less evidence than I desired to prove the bettering of the produce of very small seeds. The seeds smaller than Q were such a miserable set that I could hardly deal with them. Moreover, they were very infertile. It did, however, happen that in a few of the sets some of the Q seeds turned out very well.

If I desired to lay much stress on these experiments, I could make my case considerably stronger by going minutely into other details, including confirmatory measurements of the foliage and length of pod, but I do not care to do so.

D.

GOOD AND BAD TEMPER IN ENGLISH FAMILIES.[1]

ONE of the questions put to the compilers of the Family Records spoken of in page 72, referred to the "Character and Temperament" of the persons described. These were distributed through

[1] Reprinted after slight revision from *Fortnightly Review*, July, 1887.

three and sometimes four generations, and consisted of those who lay in the main line of descent, together with their brothers and sisters.

Among the replies, I find that much information has been incidentally included concerning what is familiarly called the "temper" of no less than 1,981 persons. As this is an adequate number to allow for many inductions, and as temper is a strongly marked characteristic in all animals; and again, as it is of social interest from the large part it plays in influencing domestic happiness for good or ill, it seemed a proper subject for investigation.

The best explanation of what I myself mean by the word "temper" will be inferred from a list of the various epithets used by the compilers of the Records, which I have interpreted as expressing one or other of its qualities or degrees. The epithets are as follows, arranged alphabetically in the two main divisions of good and bad temper:—

Good temper.—Amiable, buoyant, calm, cool, equable, forbearing, gentle, good, mild, placid, self-controlled, submissive, sunny, timid, yielding. (15 epithets in all.)

Bad temper.—Acrimonious, aggressive, arbitrary, bickering, capricious, captious, choleric, contentious, crotchety, decisive, despotic, domineering, easily offended, fiery, fits of anger, gloomy, grumpy, harsh, hasty, headstrong, huffy, impatient, imperative, impetuous, insane temper, irritable, morose, nagging, obstinate, odd-tempered, passionate, peevish, peppery, proud, pugnacious, quarrelsome, quick-tempered, scolding, short, sharp, sulky, sullen, surly, uncertain, vicious, vindictive. (46 epithets in all.)

I also grouped the epithets as well as I could, into the following five classes: 1, mild; 2, docile; 3, fretful; 4, violent; 5, masterful.

Though the number of epithets denoting the various kinds of bad temper is three times as large as that used for the good, yet the number of persons described under the one general head is about the same as that described under the other. The first set of data that I tried, gave the proportion of the good to the bad-tempered as 48 to 52; the second set as 47 to 53. There is little difference between the two sexes in the frequency of good and bad temper, but that little is in favour of the women, since about 45 men are re-

corded as good-tempered for every 55 who are bad, and conversely
55 women as good-tempered for 45 who are bad.

I will not dwell on the immense amount of unhappiness, ranging
from family discomfort down to absolute misery, or on the breaches
of friendship that must have been occasioned by the cross-grained,
sour, and savage dispositions of those who are justly labelled by
some of the severer epithets ; or on the comfort, peace, and good-
will diffused through domestic circles by those who are rightly
described by many of the epithets in the first group. We can
hardly, too, help speculating uneasily upon the terms that our own
relatives would select as most appropriate to our particular selves.
But these considerations, interesting as they are in themselves, lie
altogether outside the special purpose of this inquiry.

In order to ascertain the facts of which the above statistics are a
brief summary, I began by selecting the larger families out of my
lists, namely, those that consisted of not less than four brothers or
sisters, and by noting the persons they included who were described
as good or bad-tempered ; also the remainder about whose temper
nothing was said either one way or the other, and whom perforce I
must call neutral. I am at the same time well aware that, in some
few cases a tacit refusal to describe the temper should be interpreted as reticence in respect to what it was thought undesirable
even to touch upon.

I found that out of a total of 1,361 children, 321 were described
as good-tempered, 705 were not described at all, and 342 were
described as bad-tempered. These numbers are nearly in the proportion of 1, 2, and 1, that is to say, the good are equal in number
to the bad-tempered, and the neutral are just as numerous as the
good and bad-tempered combined.

The equality in the total records of good and bad tempers is an
emphatic testimony to the correct judgments of the compilers in the
choice of their epithets, for whenever a group has to be divided into
three classes, of which the second is called neutral, or medium, or
any other equivalent term, its nomenclature demands that it should
occupy a strictly middlemost position, an equal number of contrasted cases flanking it on either hand. If more cases were
recorded of good temper than of bad, the compilers would have laid
down the boundaries of the neutral zone unsymmetrically, too far

from the good end of the scale of temper, and too near the bad end. If the number of cases of bad temper exceeded that of the good, the error would have been in the opposite direction. But it appears, on the whole, that the compilers of the records have erred neither to the right hand nor to the left. So far, therefore, their judgments are shown to be correct.

Next as regards the proportion between the number of those who rank as neutrals to that of the good or of the bad. It was recorded as 2 to 1 ; is that the proper poportion? Whenever the nomenclature is obliged to be somewhat arbitrary, a doubtful term should be interpreted in the sense that may have the widest suitability. Now a large class of cases exist in which the interpretation of the word neutral is fixed. It is that in which the three grades of magnitude are conceived to result from the various possible combinations of two elements, one of which is positive and the other negative, such as good and bad, and which are supposed to occur on each occasion at haphazard, but in the long run with equal frequency. The number of possible combinations of the two elements is only four, and each of these must also in the long run occur with equal frequency. They are: 1, both positive; 2, the first positive, the second negative ; 3, the first negative, the second positive; 4, both negative. In the second and third of these combinations the negative counterbalances the positive, and the result is neutral. Therefore the proportions in which the several events of good, neutral, and bad would occur is as 1, 2, and 1. These proportions further commend themselves on the ground that the whole body of cases is thereby divided into two main groups, equal in number, one of which includes all neutral or medium cases, and the other all that are exceptional. Probably it was this latter view that was taken, it may be half unconsciously, by the compilers of the Records. Anyhow, their entries conform excellently to the proportions specified, and I give them credit for their practical appreciation of what seems theoretically to be the fittest standard. I speak, of course, of the Records taken as a whole ; in small groups of cases the proportion of the neutral to the rest is not so regular.

The results shown in Table I. are obtained from all my returns. It is instructive in many ways, and not least in showing to a statistical eye how much and how little value may reasonably be

attached to my materials. It was primarily intended to discover
whether any strong bias existed among the compilers to spare the
characters of their nearest relatives. In not a few cases they have
written to me, saying that their records had been drawn up with
perfect frankness, and earnestly reminding me of the importance of
not allowing their remarks to come to the knowledge of the persons
described. It is almost needless to repeat what I have published
more than once already, that I treat the Records quite confidentially.
I have left written instructions that in case of my death they should
all be destroyed unread, except where I have left a note to say that
the compiler wished them returned. In some instances I know that
the Records were compiled by a sort of family council, one of its
members acting as secretary; but I doubt much whether it often
happened that the Records were known to many of the members of
the family in their complete form. Bearing these and other considerations in mind, I thought the best test for bias would be
to divide the entries into two contrasted groups, one including
those who figured in the pedigrees as either father, mother, son, or
daughter—that is to say, the compiler and those who were very
nearly related to him—and the other including the uncles and
aunts on both sides.

TABLE 1.

DISTRIBUTION OF TEMPER IN FAMILIES (per cents.)

Relationships.	1. Mild;	2. Docile.	3. Fretful.	4. Violent.	5. Masterful	Total.	No. of cases observed.
a. Fathers and Sons	35	12	32	12	9	100	188
b. Mothers and Daughters	39	18	31	8	4	100	179
c. Uncles....................	32	13	25	18	12	100	272
d. Aunts	39	14	29	9	9	100	238
a + b. Direct line..........	74	30	63	20	13	200	367
c + d. Collaterals..........	71	27	54	27	21	200	510

	Good.	Bad Temper.			
a + b. Direct line..........	104	96		200	367
c + d. Collaterals..........	98	102		200	510

On comparing the entries, especially the summaries in the lower lines of the Table, it does not seem that the characters of near relatives are treated much more tenderly than those of the more remote. There is little indication of the compilers having been biased by affection, respect, or fear. More cases of a record being left blank when a bad temper ought to have been recorded, would probably occur in the direct line, but I do not see how this could be tested. An omission may be due to pure ignorance; indeed I find it not uncommon for compilers to know very little of some of their uncles or aunts. The Records seem to be serious and careful compositions, hardly ever used as vehicles for personal animosity, but written in much the same fair frame of mind that most people force themselves into when they write their wills.

TABLE 2.

COMBINATIONS OF TEMPER IN MARRIAGE (per cents.).

Tempers of Husbands.	A.—Observed Pairs.					B.—Haphazard Pairs.				
	Tempers of Wives.					Tempers of Wives.				
	Good.		Bad Tempers.			Good.		Bad Tempers.		
	1	2	3	4	5	1	2	3	4	5
Good 1	6	10	9	6	2	13	5	10	3	2
" 2	4	2	5	2	—	5	2	4	1	1
Bad 3	14	4	9	3	2	11	5	8	2	2
" 4	7	3	3	2	1	6	2	5	1	1
" 5	3	—	2	—	1	4	2	3	1	1
Good...........	22		24			25		21		
Bad	31		23			30		24		

The sexes are separated in the Table, to show the distribution of the five classes of temper among them severally. There is a large proportion of the violent and masterful among the men, of the fretful, the mild, and the docile among the women. On adding the entries it will be found that the proportion of those who fall within the several classes are 36 per cent. of mild-tempered, 15 per cent. of docile, 29 per cent. of fretful, 12 per cent of violent, 8 per cent. of masterful.

The importance assigned in marriage-selection to good and bad temper is an interesting question, not only from its bearing on domestic happiness, but also from the influence it may have in promoting or retarding the natural good temper of our race, assuming, as we may do for the moment, that temper is hereditary. I cannot deal with the question directly, but will give some curious facts in Table II. that throw indirect light upon it. There a comparison is made of (A) the actual frequency of marriage between persons, each of the various classes of temper, with (B) the calculated frequency according to the laws of chance, on the supposition that there had been no marriage-selection at all, but that the pairings, so far as temper is concerned, had been purely at haphazard. There are only 111 marriages in my lists in which the tempers of both parents are recorded. On the other hand, the number of possible combinations in couples of persons who belong to the five classes of temper is very large, so I make the two groups comparable by reducing both to percentages.

It will be seen that with two apparent exceptions in the upper left-hand corners of either Table (of 6 against 13, and of 10 against 5), there are no indications of predilection for, or avoidance of marriage between persons of any of the five classes, but that the figures taken from observation run as closely with those derived through calculation, as could be expected from the small number of observations. The apparent exceptions are that the percentage of mild-tempered men who marry mild-tempered women is only 6, as against 13 calculated by the laws of chance, and that those who marry docile wives are 10, as against a calculated 5. There is little difference between mildness and docility, so we may throw these entries together without much error, and then we have 6 and 10, or 16, as against 13 and 5, or 18, which is a close approximation. We may compare the frequency of marriages between persons of like temper in each of the five classes by reading the Table diagonally. They are as (6), 2, 9, 2, 1, in the observed cases, against (13), 2, 8, 1, 1, in the calculated ones; here the irregularity of the 6 and 13, which are put in brackets for distinction sake, is

conspicuous.   Elsewhere there is not the slightest indication of a
dislike in persons of similar tempers, whether mild, docile, fretful,
violent, or masterful, to marry one another.   The large initial
figures 6 and 13 catch the eye, and at a first glance impress themselves unduly on the imagination, and might lead to erroneous
speculations about mild tempered persons, perhaps that they find one
another rather insipid ; but the reasons I have given, show conclu-
sively that the recorded rarity of the marriages between mild-tempered
persons is only apparent.   Lastly, if we disregard the five smaller
classes and attend only to the main divisions of good and bad
temper, there does not appear to be much bias for, or against, the
marriage of good or bad-tempered persons in their own or into the
opposite division.

The admixture of different tempers among the brothers and
sisters of the same family is a notable fact, due to various causes
which act in different directions.   It is best to consider them before
we proceed to collect evidence and attempt its interpretation.   It
becomes clear enough, and may be now taken for granted, that the
tempers of progenitors do not readily blend in the offspring, but
that some of the children take mainly after one of them, some after
another, but with a few threads, as it were, of various ancestral
tempers woven in, which occasionally manifest themselves.   If no
other influences intervened, the tempers of the children in the same
family would on this account be almost as varied as those of their
ancestors ; and these, as we have just seen, married at haphazard,
so far as their tempers were concerned ; therefore the numbers of
good and bad children in families would be regulated by the same
laws of chance that apply to a gambling table.   But there are other
influences to be considered.   There is a well-known tendency to
family likeness among brothers and sisters, which is due, not to
the blending of ancestral peculiarities, but to the prepotence of one
of the progenitors, who stamped more than his or her fair share of
qualities upon the descendants.   It may be due also to a familiar
occurrence that deserves but has not yet received a distinctive name,
namely, where all the children are alike and yet their common
likeness cannot be traced to their progenitors.   A new variety has
come into existence through a process that affects the whole Fraternity and may result in an unusually stable variety (see Chapter III.).
The most strongly marked family type that I have personally met

with, first arose simultaneously in the three brothers of a family who transmitted their peculiarities with unusual tenacity to numerous descendants through at least four generations. Other influences act in antagonism to the foregoing; they are the events of domestic life, which instead of assimilating tempers tend to accentuate slight differences in them. Thus if some members of a family are a little submissive by nature, others who are naturally domineering are tempted to become more so. Then the acquired habit of dictation in these reacts upon the others and makes them still more submissive. In the collection I made of the histories of twins who were closely alike, it was most commonly said that one of the twins was guided by the other. I suppose that after their many childish struggles for supremacy, each finally discovered his own relative strength of character, and thenceforth the stronger developed into the leader, while the weaker contentedly subsided into the position of being led. Again, it is sometimes observed that one member of an otherwise easygoing family, discovers that he or she may exercise considerable power by adopting the habit of being persistently disagreeable whenever he or she does not get the first and best of everything. Some wives contrive to tyrannise over husbands who are mild and sensitive, who hate family scenes and dread the disgrace attending them, by holding themselves in readiness to fly into a passion whenever their wishes are withstood. They thus acquire a habit of "breaking out," to use a term familiar to the warders of female prisons and lunatic asylums; and though their relatives and connections would describe their tempers by severe epithets, yet if they had married masterful husbands their characters might have developed more favourably.

To recapitulate briefly, one set of influences tends to mix good and bad tempers in a family at haphazard; another set tends to assimilate them, so that they shall all be good or all be bad; a third set tends to divide each family into contrasted portions. We have now to ascertain the facts and learn the results of these opposing influences.

In dealing with the distribution of temper in Fraternities,[1] we

---

[1] A Fraternity consists of the brothers of a family, and of the sisters after the qualities of the latter have been transmuted to their Male Equivalents; but as no change in the Female values seems really needed, so none has been made in respect to Temper.

can only make use of those in which at least two cases of temper
are recorded; they are 146 in number. I have removed all the
cases of neutral temper, treating them as if they were non-existent,
and dealing only with the remainder that are good or bad. We
have next to eliminate the haphazard element. Beginning with
Fraternities of two persons only, either of whom is just as likely to
be good as bad tempered, there are, as we have already seen, four
possible combinations, resulting in the proportions of 1 case of both
good, 2 cases one good and one bad, and one case of both bad. I
have 42 such Fraternities, and the observed facts are that in 10
of them both are good tempered, in 20 one is good and one bad,
and in 12 both are bad tempered. Here only a trifling and untrustworthy difference is found between the observed and the
haphazard distribution, the other conditions having neutralised
each other. But when we proceed to larger Fraternities the test
becomes shrewder, and the trifling difference already observed
becomes more marked, and is at length unmistakable. Thus the
successive lines of Table III. show a continually increasing diverg-
ence between the observed and the haphazard distribution of
temper, as the Fraternities increase in size. A compendious com-

TABLE 3.

DISTRIBUTION OF TEMPER IN FRATERNITIES.

		A.—Observed.			B.—Haphazard.		
Number in each Fraternity.	Number of Fraternities.	All good-tempers.	Intermediate cases.	All bad-tempers.	All good-tempers.	Intermediate cases.	All bad-tempers.
2	42	10	20	12	10	21	11
3	55	11	15 21	8	7	20 21	7
4	29	5	6 9 8	1	2	8 12 8	2
5	6	1	0 2 1 0	2	0	1 2 2 1	0
6	14	1	0 1 3 3 2	4	0	2 4 5 4 2	0
4 to 6	49	7		7	2		2

parison is made in the bottom line of the Table by adding together the instances in which the Fraternities are from 4 to 6 in number, and in taking only those in which all the members of the Fraternity were alike in temper, whether good or bad. There are 7 + 7, or 14, observed cases of this against 2 + 2, or 4, haphazard cases, foun1 in a total of 49 Fraternities. Hence it follows that the domestic influences that tend to differentiate temper wholly fail to overcome the influences, hereditary and other, that tend to make it uniform in the same Fraternity.

As regards direct evidences of heredity of temper, we must frame our inquiries under a just sense of the sort of materials we have to depend upon. They are but coarse portraits scored with white or black, and sorted into two heaps, irrespective of the gradations of tint in the originals. The processes 1 have used in discussing the heredity of stature, eye-colour, and artistic faculty, cannot be employed in dealing with the heredity of temper. I must now renounce those refined operations and set to work with ruder tools on my rough material.

The first inquiry will be, Do good-tempered parents have, on the whole, good-tempered children, and do bad-tempered parents have bad-tempered ones? I have 43 cases where both parents are recorded as good-tempered, and 25 where they were both bad-tempered. Out of the children of the former, 30 per cent. were good-tempered and 10 per cent. bad; out of the latter, 4 per cent. were good and 52 per cent, bad-tempered. This is emphatic testimony to the heredity of temper. I have worked out the other less contrasted combinations of parental temper, but the results are hardly worth giving. There is also much variability in the proportions of the neutral cases.

I then attempted, with still more success, to answer the converse question, Do good-tempered Fraternities have, on the whole, good-tempered ancestors, and bad tempered Fraternities bad-tempered ones? After some consideration of the materials, I defined—rightly or wrongly—a good-tempered Fraternity as one in which at least two members were good-tempered and none were bad, and a bad-tempered Fraternity as one in which at least two members were bad-tempered, whether or no any cases of good temper were said to be associated with them. Then, as regards the ancestors, I thought

by far the most trustworthy group was that which consisted of
the two parents and of the uncles and aunts on both sides. I
have thus 46 good-tempered Fraternities with an aggregate of 333
parents, uncles, and aunts ; and 71 bad-tempered, with 633 parents,
uncles, and aunts. In the former group, 26 per cent. were good
tempered and 18 bad ; in the latter group, 18 were good-tempered
and 29 were bad, the remainder being neutral. These results are
almost the exact counterparts of one another, so I seem to have
made good hits in framing the definitions. More briefly, we may
say that when the Fraternity is good-tempered as above defined,
the number of good-tempered parents, uncles, and aunts, exceeds
that of the bad-tempered in the proportion of 3 to 2 ; and that
when the Fraternity is bad-tempered, the proportions are exactly
reversed.

I have attempted in other ways to work out the statistics of
hereditary tempers, but none proved to be of sufficient value for
publication. I can trace no prepotency of one sex over the other
in transmitting their tempers to their children. I find clear
indications of strains of bad temper clinging to families for three
generations, but I cannot succeed in putting them into a numerical
form.

It must not be thought that I have wished to deal with temper
as if it were an unchangeable characteristic, or to assign more
trustworthiness to my material than it deserves. Both these
views have been discussed ; they are again alluded to to show
that they are not dismissed from my mind, and partly to give the
opportunity of adding a very few further remarks.

Persons highly respected for social and public qualities may be
well-known to their relatives as having sharp tempers under strong
but insecure control, so that they "flare up" now and then. I
have heard the remark that those who are over-suave in ordinary
demeanour have often vile tempers. If this be the case—and I
have some evidence of its truth—I suppose they are painfully
conscious of their infirmity, and through habitual endeavours to
subdue it, have insensibly acquired an exaggerated suavity at the
times when their temper is unprovoked. Illness, too, has much
influence in affecting the temper. Thus I sometimes come across
entries to the effect of, " not naturally ill-tempered, but peevish

through illness." Overwork and worry will make even mild-tempered men exceedingly touchy and cross.

The accurate discernment and designation of character is almost beyond the reach of any one, but, on the other hand, a rough estimate and a fair description of its prominent features is easily obtainable; and it seems to me that the testimony of a member of a family who has seen and observed a person in his unguarded moments and under very varied circumstances for many years, is a verdict deserving of much confidence. I shall have fulfilled my object in writing this paper if it leaves a clear impression of the great range and variety of temper among persons of both sexes in the upper and middle classes of English society; of its disregard in Marriage Selection; of the great admixture of its good and bad varieties in the same family; and of its being, nevertheless, as hereditary as any other quality. Also, that although it exerts an immense influence for good or ill on domestic happiness, it seems that good temper has not been especially looked for, nor ill temper especially shunned, as it ought to be in marriage-selection.

E.

THE GEOMETRIC MEAN, IN VITAL AND SOCIAL STATISTICS.[1]

My purpose is to show that an assumption which lies at the basis of the well-known law of "Frequency of Error" is incorrect when applied to many groups of vital and social phenomena, although that law has been applied to them by statisticians with partial success. Next, I will point out the correct hypothesis upon which a Law of Error suitable to these cases ought to be calculated; and subsequently I will communicate a memoir by Mr. (now Dr.) Donald Macalister, who, at my suggestion, has mathematically investigated the subject.

The assumption to which I refer is, that errors in excess or in deficiency of the truth are equally probable; or conversely, that if two fallible measurements have been made of the same object, their

[1] Reprinted, with slight revision, from the *Proceedings of the Royal Society*, No. 198, 1879.

arithmetical mean is more likely to be the true measurement than any other quantity that can be named.

This assumption cannot be justified in vital phenomena. For example, suppose we endeavour to match a tint; Weber's law, in its approximative and simplest form, of Sensation varying as the logarithm of the Stimulus, tells us that a series of tints, in which the quantities of white scattered on a black ground are as 1, 2, 4, 8, 16, 32, &c., will appear to the eye to be separated by equal intervals of tint. Therefore, in matching a grey that contains 8 portions of white, we are just as likely to err by selecting one that has 16 portions as one that has 4 portions. In the first case there would be an error in excess, of 8 units of absolute tint; in the second there would be an error in deficiency, of 4. Therefore, an error of the same magnitude in excess or in deficiency is not equally probable in the judgment of tints by the eye. Conversely, if two persons, who are equally good judges, describe their impressions of a certain tint, and one says that it contains 4 portions of white and the other that it contains 16 portions, the most reasonable conclusion is that it really contains 8 portions. The arithmetic mean of the two estimates is 10, which is *not* the most probable value; it is the geometric mean 8, (4 : 8 : : 8 : 16), which is the most probable.

Precisely the same condition characterises every determination by each of the senses; for example, in judging of the weight of bodies or of their temperatures, of the loudness and of the pitches of tones, and of estimates of lengths and distances *as wholes*. Thus, three rods of the lengths a, b, c, when taken successively in the hand, appear to differ by equal intervals when $a : b : : b : c$, and not when $a - b = b - c$. In all physiological phenomena, where there is on the one hand a stimulus and on the other a response to that stimulus Weber's or some other geometric law may be assumed to prevail in other words, the true mean is geometric rather than arithmetic.

The geometric mean appears to be equally applicable to the majority of the influences, which, combined with those of purely vital phenomena, give rise to the events with which sociology deals. It is difficult to find terms sufficiently general to apply to the varied topics of sociology, but there are two categories which are of common occurrence in which the geometric mean is certainly appropriate. The one is increase, as exemplified by the growth of population, where an

already large nation tends to receive larger accessions than a small
one under similar circumstances, or when a capital employed in a
business increases in proportion to its size. The other category is
the influences of circumstances or of "milieux" as they are often
called, such as a period of plenty in which a larger field or a larger
business yields a greater excess over its mean yield than a smaller
one. Most of the causes of those differences with which sociology are
concerned, and which are not purely vital phenomena, such as those
previously discussed, may be classified under one or other of these
two categories, or under such as are in principle almost the same.
In short, sociological phenomena, like vital phenomena are, as a
general rule, subject to the condition of the geometric mean.

   The ordinary law of Frequency of Error, based on the arithmetic
mean, corresponds, no doubt, sufficiently well with the observed facts
of vital and social phenomena, to be very serviceable to statisticians,
but it is far from satisfying their wants, and it may lead to absurdity
when applied to wide deviations. It asserts that deviations in excess
must be balanced by deviations of equal magnitude in deficiency;
therefore, if the former be greater than the mean itself, the latter
must be less than zero, that is, must be negative. This is an impossibility in many cases, to which the law is nevertheless applied by statisticians with no small success, so long as they are content to confine
its application within a narrow range of deviation. Thus, in respect
of Stature, the law is very correct in respect to ordinary measurements, although it asserts that the existence of giants, whose height
is more than double the mean height of their race, implies the possibility of the existence of dwarfs, whose stature is less than nothing
at all.

   It is therefore an object not only of theoretical interest but of
practical use, to thoroughly investigate a Law of Error, based on the
geometric mean, even though some of the expected results may
perhaps be apparent at first sight.   With this view I placed the foregoing remarks in Mr. Donald Macalister's hands, who contributed
a memoir that will be found in the *Proc. Royal Soc.*, No. 198, 1879,
following my own. It should be referred to by such mathematicians
as may read this book.

F.

PROBABLE EXTINCTION OF FAMILIES.[1]

THE decay of the families of men who occupied conspicuous positions in past times has been a subject of frequent remark, and has given rise to various conjectures. It is not only the families of men of genius or those of the aristocracy who tend to perish, but it is those of all with whom history deals, in any way, even such men as the burgesses of towns, concerning whom Mr. Doubleday has inquired and written. The instances are very numerous in which surnames that were once common have since become scarce or have wholly disappeared. The tendency is universal, and, in explanation of it, the conclusion has been hastily drawn that a rise in physical comfort and intellectual capacity is necessarily accompanied by diminution in "fertility"—using that phrase in its widest sense and reckoning abstinence from marriage as one cause of sterility. If that conclusion be true, our population is chiefly maintained through the "proletariat," and thus a large element of degradation is inseparably connected with those other elements which tend to ameliorate the race. On the other hand, M. Alphonse de Candolle has directed attention to the fact that, by the ordinary law of chances, a large proportion of families are continually dying out, and it evidently follows that, until we know what that proportion is, we cannot estimate whether any observed diminution of surnames among the families whose history we can trace, is or is not a sign of their diminished "fertility." I give extracts from M. De Candolle's work in a foot-note,[2] and may add that, although I have not hitherto published anything on the matter, I took considerable pains some years ago to obtain numerical results in respect to this

[1] Reprinted, with slight revision, from the *Journ. Anthropol. Inst.* 1888.

[2] "Au milieu des renseignements précis et des opinions très-sensées de MM. Benoiston de Châteauneuf, Galton, et autres statisticiens, je n'ai pas rencontré la réflexion bien importante qu'ils auraient dû faire de l'extinction *inévitable* des noms de famille. Évidemment tous les noms doivent s'éteindre .... Un mathématicien pourrait calculer comment la réduction des noms ou titres aurait lieu, d'après la probabilité des naissances toutes féminines ou toutes masculines ou mélangées et la probabilité d'absence de naissances dans un couple quelconque," &c.—ALPHONSE DE CANDOLLE, *Histoire des Sciences et des Savants*, 1873.

R

very problem. I made certain very simple and not very inaccurate suppositions concerning average fertility, and I worked to the nearest integer, starting with 10,000 persons, but the computation became intolerably tedious after a few steps, and I had to abandon it. The Rev. H. W. Watson kindly, at my request, took the problem in hand, and his results form the subject of the following paper. They do not give what can properly be called a general solution, but they do give certain general results. They show (1) how to compute, though with great labour, any special case; (2) a remarkably easy way of computing those special cases in which the law of fertility approximates to a certain specified form; and (3) how all surnames tend to disappear.

The form in which I originally stated the problem is as follows. I purposely limited it in the hope that its solution might be more practicable if unnecessary generalities were excluded :—

A large nation, of whom we will only concern ourselves with the adult males, N in number, and who each bear separate surnames, colonise a district. Their law of population is such that, in each generation, a_0 per cent. of the adult males have no male children who reach adult life; a_1 have one such male child; a_2 have two; and so on up to a_5, who have five. Find (1) what proportion of the surnames will have become extinct after r generations; and (2) how many instances there will be of the same surname being held by m persons.

Discussion of the problem by the Rev. H. W. Watson, D.Sc., F.R.S., *formerly Fellow of Trinity College, Cambridge.*

Suppose that at any instant all the adult males of a large nation have different surnames, it is required to find how many of these surnames will have disappeared in a given number of generations upon any hypothesis, to be determined by statistical investigations, of the law of male population.

Let, therefore, a_0 be the percentage of males in any generation who have no sons reaching adult life, let a_1 be the percentage that have one such son, a_2 the percentage that have two, and so on up to a_q, the percentage that have q such sons, q being so large that it is not worth while to consider the chance of any man having more than q adult sons—our first hypothesis will be that the numbers

a_0, a_1, a_2, etc., remain the same in each succeeding generation. We shall also, in what follows, neglect the overlapping of generations—that is to say, we shall treat the problem as if all the sons born to any man in any generation came into being at one birth, and as if every man's sons were born and died at the same time. Of course it cannot be asserted that these assumptions are correct. Very probably accurate statistics would discover variations in the values of a_0, a_1, etc., as the nation progressed or retrograded ; but it is not at all likely that this variation is so rapid as seriously to vitiate any general conclusions arrived at on the assumption of the values remaining the same through many successive generations. It is obvious also that the generations must overlap, and the neglect to take account of this fact is equivalent to saying, that at any given time we leave out of consideration those male descendants, of any original ancestor who are more than a certain average number of generations removed from him, and compensate for this by giving credit for such male descendants, not yet come into being, as are not more than that same average number of generations removed from the original ancestors.

Let then $\frac{a_0}{100}$, $\frac{a_1}{100}$, $\frac{a_2}{100}$, etc., up to $\frac{a_q}{100}$ be denoted by the symbols t_0, t_1, t_2, etc., up to t_q, in other words, let t_0, t_1, etc., be the chances in the first and each succeeding generation of any individual man, in any generation, having no son, one son, two sons, and so on, who reach adult life. Let N be the original number of distinct surnames, and let $_rm_s$ be the fraction of N which indicates the number of such surnames with s representatives in the rth generation.

Now, if any surname have p representatives in any generation, it follows from the ordinary theory of chances that the chance of that same surname having s representatives in the next succeeding generation is the coefficient of x^s in the expansion of the multinomial

$$(t_0 + t_1x + t_2x^2 +, \text{ etc. } + t_qx^q)^p$$

Let then the expression $t_0 + t_1x + t_2x^2 +$ etc. $+ t_qx^q$ be represented by the symbol T.

Then since, by the assumption already made, the number of surnames with no representative in the $r-1$th generation is $_{r-1}m_0$ N, the number with one representative $_{r-1}m_1.N$, the number with two $_{r-1}m_2$. N and so on, it follows, from what we last stated, that the number of surnames with s representatives in the rth generation must be the coefficient of x^s in the expression

$$\left\{ \ _{r-1}m_0 + _{r-1}m_1 T + _{r-1}m_2 T^2 + \text{ etc. } + _{r-1}m_{qr-1}T^{qr-1} \ \right\} N$$

If, therefore, the coefficient of N in this expression be denoted by $f_r(x)$ it follows that $_{r-1}m_1$, $_{r-1}m_2$ and so on, are the coefficients of x, x^2 and so on, in the expression $f_{r-1}(x)$.

If, therefore, a series of functions be found such that

$$f_1(x) = t_0 + t_1 x + \text{ etc. } + t_q x^q \text{ and } f_r(x) = f_{r-1}(t_0 + t_1 x \text{ etc. } + t^q x^q)$$

then the proportional number of groups of surnames with s representatives in the rth generation will be the coefficient of x^s in $f_r(x)$ and the actual number of such surnames will be found by multiplying this coefficient by N. The number of surnames unrepresented or become extinct in the rth generation will be found by multiplying the term independent of x in $f_r(x)$ by the number N.

The determination, therefore, of the rapidity of extinction of surnames, when the statistical data, t_0, t_1, etc., are given, is reduced to the mechanical, but generally laborious process of successive substitution of $t_0 + t_1 x + t_2 x^2 + $ etc., for x in successively determined values of $f_r(x)$, and no further progress can be made with the problem until these statistical data are fixed; the following illustrations of the application of our formula are, however, not without interest.

(1) The very simplest case by which the formula can be illustrated is when $q = 2$ and t_0, t_1, t_2 are each equal to $\frac{1}{3}$.

Here $f_1(x) = \dfrac{1+x+x^2}{3} f_2(x) = \dfrac{1}{3} \left\{ 1 + \dfrac{1}{3}(1+x+x^2) + \dfrac{1}{9} \left. 1 + x + x^2)^2 \right\}^2 \right.$

and so on.

Making the successive substitutions, we obtain

$$f_2(x) = \frac{1}{3} \left\{ \frac{13}{9} + \frac{5x}{9} + \frac{6x^3}{9} + \frac{2x}{9} + \frac{x}{9} \right\}$$

$$f_3(x) = \frac{1249}{2187} + \frac{265x}{2187} + \frac{343x^3}{2187} + \frac{166x^3}{2187} + \frac{109x^4}{2187} + \frac{34x^5}{2187} + \frac{16x^6}{2187} + \frac{4x^7}{2187} + \frac{x^8}{2187}$$

$$f_4(x) = \cdot 63183 + \cdot 08306x + \cdot 10635x^2 + \cdot 07804x^3 + \cdot 06489x^4 + \cdot 05443x^5 + \cdot 01437x^6$$
$$+ \cdot 01692x^7 + \cdot 01144x^8 + \cdot 00367x^9 + \cdot 00104x^{10} + \cdot 00015x^{11} + \cdot 00005x^{12}$$
$$+ \cdot 00001x^{13} + \cdot 00000x^{14} + \cdot 00000x^{15} + \cdot 00000x^{16}$$

and the constant term in $f_5\,(x)$ or $_5m_0$ is therefore

$$6\cdot3183 + \frac{\cdot08306}{3} + \frac{\cdot10635}{9} + \frac{\cdot07804}{27} + \frac{\cdot06489}{81} + \frac{\cdot05443}{243} + \frac{\cdot01437}{729} + \frac{\cdot01692}{2187} + \frac{\cdot01144}{6561}$$

$$+ \frac{\cdot00367}{19683} + \frac{\cdot00104}{59049} + \frac{\cdot00015}{177147} +$$

The value of which to five places of decimals is $\cdot67528$.

The constant terms, therefore, in f_1, f_2 up to f_5 when reduced to decimals, are in this case $\cdot33333$, $\cdot48148$, $\cdot57110$, $\cdot64113$, and $\cdot65628$ respectively. That is to say, out of a million surnames at starting, there have disappeared in the course of one, two, etc., up to five generations, 333333, 481480, 571100, 641130, and 675280 respectively.

The disappearances are much more rapid in the earlier than in the later generations. Three hundred thousand disappear in the first generation, one hundred and fifty thousand more in the second, and so on, while in passing from the fourth to the fifth, not more than thirty thousand surnames disappear.

All this time the male population remains constant. For it is evident that the male population of any generation is to be found by multiplying that of the preceding generation, by $t_1 + 2t_2$, and this quantity is in the present case equal to one.

If axes Ox and Oy be drawn, and equal distances along Ox represent generations from starting, while two distances are marked along every ordinate, the one representing the total male population in any generation, and the other the number of remaining surnames in that generation, of the two curves passing through the extremities of these ordinates, the *population* curve will, in this case, be a straight line parallel to Ox, while the *surname* curve will intersect the population curve on the axis of y, will proceed always convex to the axis of x, and will have the positive part of that axis for an asymptote.

The case just discussed illustrates the use to be made of the general formula, as well as the labour of successive substitutions, when the expressions $f_1\,(x)$ does not follow some assigned law. The calculation may be infinitely simplified when such a law can be found; especially if that law be the expansion of a binomial, and only the extinctions are required.

For example, suppose that the terms of the expression $t_0 + t_1 x +$ etc. $+ t_q x_q$ are proportional to the terms of the expanded binomial

$(a+bx)^q$ *i.e.* suppose that $t_0 = \dfrac{a^q}{(a+b)^q}$, $t_1 = q\dfrac{a^{q+1}b}{(a+b)^q}$ and so on.

Here $f_1(x) = \dfrac{(a+bx)^q}{(a+b)^q}$ and $_1m_0 = \dfrac{a^q}{(a+b)^q}$

$$f_2(x) = \frac{1}{(a+b)^q}\left\{ a+b\frac{(a+bx)^q}{(a+b)^q} \right\}^q$$

$$_2m_0 = \frac{1}{(a+b)^q}\left\{ a+b_1m_0 \right\}^q$$

Generally $_rm_0 = \dfrac{1}{(a+b)^q}\left\{ a+b_{r-1}m_0 \right\}^q = \dfrac{b_q}{(a+b)_q}\left\{ \dfrac{a}{b}+_{r-1}m_0 \right\}^q$

If, therefore, we wish to find the number of extinctions in any generation, we have only to take the number in the preceding generation, add it to the constant fraction $\dfrac{a}{b}$, raise the sum to the power of q, and multiply by $\dfrac{b^q}{(a+b)^q}$

With the aid of a table of logarithms, all this may be effected for a great number of generations in a very few minutes. It is by no means unlikely that when the true statistical data t_0, t_1, etc., t_q are ascertained, values of a, b, and q may be found, which shall render the terms of the expansion $(a+bx)^q$ approximately proportionate to the terms of $f_1(x)$. If this can be done, we may *approximate* to the determination of the rapidity of extinction with very great ease, for any number of generations, however great.

For example, it does not seem very unlikely that the value of q might be 5, while t_0, t_1...t_q might be ·237, ·396, ·264, ·088, ·014, ·001, or nearly, $\frac{1}{4}$, $\frac{1}{3}$, $\frac{7}{24}$, $\frac{1}{25}$, $\frac{1}{138}$, and $\frac{1}{1000}$.

Should that be the case, we have, $f_1(x) = \dfrac{(3+x)^5}{4^5}$ $_1m_0 = \dfrac{3^5}{4^5}$

and generally $_rm_0 = \dfrac{1}{4^5}\left\{ 3+_{r-1}m_0 \right\}^{·5}$

Thus we easily get for the number of extinctions in the first ten generations respectively.

·237, ·346, ·410, ·450, ·477, ·496, ·510, ·520, ·527, ·533.

We observe the same law noticed above in the case of $\dfrac{1+x+x^2}{3}$ viz., that while 237 names out of a thousand disappear in the first step, and an additional 109 names in the second step, there are only 27 disappearances in the fifth step, and only six disappearances in the tenth step.

If the curves of surnames and of population were drawn from this case, the former would resemble the corresponding curve in the case last mentioned, while the latter would be a curve whose distance from the axis of x increased indefinitely, inasmuch as the expression

$$t_1 + 2t_2 + 3t_3 + 4t_4 + 5t_5$$

is greater than one.

Whenever $f_1(x)$ can be represented by a binomial, as above suggested, we get the equation

$$_r m_0 = \frac{1}{(a+b)^q} \left\{ a + b_{r-1}m \right\}^q$$

whence it follows that as r increases indefinitely the value of $_r m_0$ approaches indefinitely to the value y where

$$y = \frac{1}{(a+b)} \left\{ a + by \right\}$$

that is where $y = 1$.

All the surnames, therefore, tend to extinction in an indefinite time, and this result might have been anticipated generally, for a surname once lost can never be recovered, and there is an additional chance of loss in every successive generation. This result must not be confounded with that of the extinction of the male population ; for in every binomial case where q is greater than 2, we have $t_1 + 2t_2 +$ etc. $+ qt_q > 1$, and, therefore an indefinite increase of male population.

The true interpretation is that each of the quantities, $_r m_1$, $_r m_2$, etc., tends to become zero, as r is indefinitely increased, but that it does not follow that the product of each by the infinitely large number N is also zero.

As, therefore, time proceeds indefinitely, the number of surnames extinguished becomes a number of the *same order of magnitude* as the total number at first starting in N, while the number of surnames represented by one, two, three, etc., representatives is some infinitely smaller but finite number. When the finite numbers are multiplied by the corresponding number of representatives, sometimes infinite in number, and the products added together, the sum will generally exceed the original number N. In point of fact, just as in the cases calculated above to five generations, we had a continual, and indeed at first, a rapid extinction of surnames, combined in the one case with a stationary, and in the other case an increasing population, so is it when the number of generations is increased indefinitely. We have a continual extinction of surnames going on, combined with constancy, or increase of population, as the case may be, until at length the number of surnames remaining is absolutely insensible, as compared with the number at starting; but the total number of representatives of those remaining surnames is infinitely greater than the original number.

We are not in a position to assert from *actual calculation* that a corresponding result is true for every form of $f_1(x)$, but the reasonable inference is that such is the case, seeing that it holds whenever $f_1(x)$ may be compared with $\dfrac{(a+bx)_q}{(a+b)_q}$ whatever a, b, or q may be.

G.

ORDERLY ARRANGEMENT OF HEREDITARY DATA.

THERE are many methods both of drawing pedigrees and of describing kinship, but for my own purposes I still prefer those that I designed myself. The chief requirements that have to be fulfilled are compactness, an orderly and natural arrangement, and clearly intelligible symbols.

Nomenclature.—A symbol ought to be suggestive, consequently the initial letter of a word is commonly used for the purpose. But this practice would lead to singular complications in symbolizing the various ranks of kinship, and it must be applied sparingly. The letter F is equally likely to suggest any one of the three very different words of Father, Female, and Fraternal. The letter M suggests both Mother and Male; S would do equally for Son and for Sister. Whether they are English, French, or German words, much the same complexity prevails. The system employed in *Hereditary Genius* had the merit of brevity, but was apt to cause mistake; it was awkward in manuscript and difficult to the printer, and I have now abandoned it in favour of the method employed in the *Records of Family Faculties.* This will now be briefly described again. Each kinsman can be described in two ways, either by letters or by a number. In ordinary cases both the letter and number are intended to be used simultaneously, thus FF.8. means the Father's Father of the person described, though either FF or 8, standing by themselves, would have the same meaning. The double nomenclature has great practical advantages. It is a check against mistake and makes reference and orderly arrangement easy.

As regards the letters, F stands for Father and M for Mother, whenever no letter succeeds them ; otherwise they stand for Father's and for Mother's respectively. Thus F is Father; FM is Father's Mother ; FMF is Father's Mother's Father.

As regards the principle upon which the numbers are assigned, arithmeticians will understand me when I say that it is in accordance with the binary system of notation, which runs parallel to the binary distribution of the successive ranks of ancestry, as two parents, four grandparents, eight great-grandparents, and so on. The "subject" of the pedigree is of generation 0 ; that of his parents, of generation 1; that of his grandparents, of generation 2, &c. This is clearly shown in the following table :—

Kinship.	Order of Generation.	Numerical Values							
		in Binary Notation.				in Decimal Notation.			
Child.............	0	1				1			
Parents	1	10		11		2		3	
Gr. Par............	2	100	101	110	111	4	5	6	7
G. Gr. Par.	3	1000	1010	1100	1110	8	10	12	14
		1001	1011	1100	1111	9	11	13	15

All the male ancestry are thus described by even numbers and the female ancestry by odd ones. They run as follows :—

F, 2.		M, 3.	
FF, 4.	FM, 5.	MF, 6.	MM, 7.
FFF, 8.	FMF, 10.	MFF, 12.	MMF, 14.
FFM, 9.	FMM, 11.	MFM, 13.	MMM, 15.

It will be observed that the double of the number of any ancestor
is that of his or her Father; and that the double of the number
plus 1 is that of his or her Mother; thus FM 5 has for her father
FMF 10, and for her mother FMM 11.

When the word Brother or Sister has to be abbreviated it is safer
not to be too stingy in assigning letters, but to write *br*, *sr*, and in
the plural *brs*, *srs*; also for the long phrase of "brothers and sisters,"
to write *brss*.

All these symbols are brief enough to save a great deal of space,
and they are perfectly explicit. When such a phrase has to be
expressed as "the Fraternity of whom FF *is* one" I write in my
own notes simply FF', but there has been no occasion to adopt this
symbol in the present book.

I have not satisfied myself as to any system for expressing
descendants. Theoretically, the above binary system admits of
extension by the use of negative indices, but the practical application of the idea seems cumbrous.

We and the French sadly want a word that the Germans possess
to stand for Brothers and Sisters. Fraternity refers properly to the
brothers only, but its use has been legitimately extended here to
mean the brothers and the sisters, after the qualities of the latter
have been reduced to their male equivalents. The Greek word
adelphic would do for an adjective.

Pedigrees.—The method employed in the *Record of Family Faculties*
for entering all the facts concerning each kinsman in a methodical
manner was fully described in that book, and could not easily be
epitomised here; but a description of the method in which the
manuscript extracts from the records have been made for my own
use will be of service to others when epitomising their own family
characteristics. It will be sufficient to describe the quarto books
that contain the medical extracts. Each page is ten and a half inches
high and eight and a half wide, and the two pages, 252, 253, that are

SCHEDULE.

JAMES LIPHOOK.

Father's Father's Father and his fraternity.

Father's Father's Mother and her fraternity.

Father's Father and his fraternity.

Spare space.

Father's Mother's Father and his fraternity.

Father's Mother's Mother and her fraternity.

Father's Mother and her fraternity.

Spare space.

Father and his fraternity.

Spare space.

JAMES LIPHOOK.

Mother's Father's Father and his fraternity.

Mother's Father's Mother and her fraternity.

Mother's Father and his fraternity.

Spare space.

Mother's Mother's Father and his fraternity.

Mother's Mother's Mother and her fraternity.

Mother's Mother and her fraternity.

Mother and her fraternity.

Spare space.

Children.

Example A.

| Father's name . . . . JAMES GLADDING. |
| Mother's maiden name . . . . MARY CLAREMONT. |

Initials.	Kin.	Principal illnesses and cause of death.	Age at death.
J. G.	Father	Bad rheum. fever ; agonising headaches ; diarrhœa ; bronchitis ; pleurisy . . *Heart disease*	54
R. G.	bro.	Rheum. gout . . . . *Apoplexy*	56
W. G.	bro.	Good health except gout ; paralysis later *Apoplexy*	83
F. L.	sis.	Rheum. fever and rheum. gout . . . *Apoplexy*	78
C. G.	sis.	Delicate . . . . . . (inoculated) *Small pox*	
M. G.	Mother	Tendency to lung disease ; biliousness ; frequent heart attacks . *Heart disease and dropsy*	63
A. C.	bro.	Good health . . . . *Accident*	46
W. C.	bro.	Led a wild life . . . . . . *Premature old age*	62
E. C.	bro.	Always delicate . . . . *Consumption*	19
F. R.	sis.	Small-pox three times . . . . *General failure*	85
R. N.	sis.	Bilious ; weak health . . . . *Cancer*	50
L. O.	sis.	. . . . *Fever*	21
M. G.	bro.	Inflam. lungs ; rheum. fever . . *Heart disease*	17
K. G.	bro.	Debility ; heart disease ; colds . *Consumption*	40
G. L.	sis.	Bad headaches ; coughs ; weak spine ; hysteria ; apoplexy . . . . *Paralysis*	50
F. S.	sis.	Bad colds ; inflam. lungs ; hysteria	living
R. F.	sis.	Infantile paralysis ; colds ; nervous depression .	living
L. G.	sis.	Inflam. brain, also lungs ; neuralgia ; nervous fever	living

(Space left for remarks.)

EXAMPLE B.

Father's name . . . . . . JULIUS FITZROY.
Mother's maiden name . . . AMELIA MERRYWEATHER.

Initials.	Kin.	Principal illnesses and cause of death.	Age at death.
R. F.	Father	Gouty Habit . . *Weak Heart and Congest. Liver*	73
L. F.¡	bro.	. . . . *Gout and Decay*	88
A: G. F.	bro.	. . . . *Accident*	48
W. F.	bro.	. . . . *Typhoid*	16
	Mother	Gall stones . . . . *Internal Malady* (?) *Cancer*	55
P. M.	bro.	. . . . *Paralysis*	86
A. M.	bro.	. . . . *Paralysis*	85
N. M.	bro.	Still living.	
R. B.	sis.	. . . . *Consumption*	33
C. M.	sis.	. . . . *Rheum. in Head*	88
F. L.	sis.	. . . . *Softening of Brain*	76
		1 died an infant.	
G. F.	bro.	Gout: tendency to mesenteric disease ; eruptive disorders . . . *Blood poisoning after a cut*	46
H. F.	bro.	Liver deranged ; bad headaches ; once supposed consumptive . . . . . *Gradual Paralysis*	45
S. T. F.	bro.	Eruptive disorder; mesentery disease ; inflammation of liver . . *Inflammation of Lungs*	42
H. G.	sis.	Eruptive disorder ; liver . . . *Inflam. of Lungs*	47
H. B. R.	sis.	Delicate ; tend. to mesent. disease . *Consumption*	29
N. F.	sis.	Colds ; liver disorder . . . . . . *Consumption*	30
E. L. F.	sis.	Mesenteric disease ; grandular swellings . *Atrophy*	16
		2 died infants.	

(Space left for remarks.)

found wherever the book is opened, relate to the same family. The open book is ruled so as to resemble the accompanying schedule, which is drawn on a reduced scale on page 251. The printing within the compartments of the schedule does not appear in the MS. books, it is inserted here merely to show to whom each compartment refers. It will be seen that the paternal ancestry are described in the left page, the maternal in the right. The method of arrangement is quite orderly, but not altogether uniform. To avoid an unpleasing arrangement like a tree with branches, and which is very wasteful of space, each grandparent and his own two parents are arranged in a set of three compartments one above the other. There are, of course, four grandparents and therefore four such sets in the schedule. Reference to the examples A and B pages 252 and 253 will show how these compartments are filled up. The rest of the Schedule explains itself. The children of the pedigree are written below the compartment assigned to the mother and her brothers and sisters; the spare spaces are of much occasional service, to receive the overflow from some of the already filled compartments as well as for notes. It is astonishing how much can be got into such a schedule by writing on ruled paper with the lines one-sixth of an inch apart, which is not too close for use. Of course the writing must be small, but it may be bold, and the figures should be written very distinctly.

For a less ambitious attempt, including the grandparents and their fraternity, but not going further back, the left-hand page would suffice, placing "Children" where "Father" now stands, "Father's Father" for "Father," and so on throughout.

INDEX.

INDEX.

THE END.

RICHARD CLAY AND SONS, LIMITED, LONDON AND BUNGAY.

Works by the Same Author.

METEOROGRAPHICA; or, Methods of

Mapping the Weather. Illustrated by upwards of 600 Printed and Lithographed Diagrams. 4to. 9s.

ENGLISH MEN OF SCIENCE: Their

Nature and Nurture. 8vo. 8s. 6d.

INQUIRIES INTO HUMAN FACULTY AND ITS DEVELOPMENT.

With Illustrations and Coloured and Plain Plates. Demy 8vo. 16s.

RECORD OF FAMILY FACULTIES.

Consisting of Tabular Forms and Directions for Entering Data, with an Explanatory Preface. 4to. 2s. 6d.

LIFE HISTORY ALBUM;

Being a Personal Note-book, combining the chief advantages of a Diary, Photograph Album, a Register of Height, Weight, and other Anthropometrical Observations, and a Record of Illnesses. Containing Tabular Forms, Charts, and Explanations especially designed for popular use.

Prepared by the direction of the Collective Investigation Committee of the British Medical Association, and Edited by FRANCIS GALTON, F.R.S., Chairman of the Life History Sub-Committee.

4to. 3s. 6d. Or, with Cards of Wools for Testing Colour Vision, 4s. 6d.

MACMILLAN & CO., LONDON.